The Boater's Weather Guide

C O R N E L L   B O A T E R S   L I B R A R Y

The Boater's Weather Guide

BY

MARGARET WILLIAMS

Cornell Maritime Press

Centreville, Maryland

Library of Congress Cataloging-in-Publication Data

Williams, Margaret, 1930-
The boater's Weather Guide / by Margaret Williams.—1st ed.
p. cm.—(Cornell boaters library)
Includes index.
ISBN 0-87033-417-4 (paper) :
1. Meteorology, Maritime. 2. Boats and boating. I. Title.
II. Series.
QC994.W49 1990 90-55451
623.88—dc20 CIP

Manufactured in the United States of America
First edition, 1990; second printing, 1996

Contents

PART II: WEATHER AND THE BOATER

Acknowledgments

Special thanks to the many people at the National Oceanographic and Atmospheric Administration for providing the photographs and synoptic charts, to C. Wayne Uptergrove for his review of the manuscript and his helpful suggestions, and to my husband for his patience and assistance in the preparation of this book.

The Boater's Weather Guide

Introduction

"It's going to be another hot one," announced the weatherman. "Temperatures in downtown Seattle will climb into the nineties again today. But it will be gorgeous—clear and sunny. So enjoy the outdoors this weekend."

We switched off the radio and tuned in the VHF marine weather channel. The story there was different: "Offshore winds are southwesterly 25 to 40 knots. . . ." We had been tied up at the guest dock in Port Angeles for nearly a week, waiting for the weather to change, but the forecast was always the same. Beating the 50-odd miles out of the Straits in high winds to the Pacific, and then beating our way offshore across an ocean with the attendant heavy seas, was not our idea of a pleasant voyage. Other boaters did not share our trepidation. They untied the docklines, waved a cheerful good-bye, and left—only to return a few hours later, cold and scared, with broken headstays, tattered sails, and a variety of other damage.

Many of us are city dwellers, only minimally affected by weather in our daily lives. On land, barring hurricanes or floods, the roads we travel remain the same regardless of weather. But in the marine environment, it is weather that determines the state of the waters we must negotiate. Once we cast off the docklines or slide a boat down the launching ramp, we are at the mercy of wind and seas. It matters little if the craft is powered by oars, sail, or motor, or whether the water to be navigated is a lake, a river, an ocean, or an inlet. Every skipper, before leaving safe harbor, must think: What is the wind strength and direction at this moment? What is it likely to be later in the day? Are conditions safe for my boat?

Whether racing or cruising, every skipper should be his or her own weather forecaster. The racing sailor must be able to keep finding wind or be forced to sit with drooping sails, wondering why the rest of the fleet is sailing smartly by. The predicted log racer must factor in the expected wind directions and strengths along the course in order to win. The weekend cruiser must consider whether a sudden wind shift will make a tranquil anchorage inhospitable. And those just out for a relaxing time away from phones and shoreside pressures certainly do not want to be caught by a frightening storm. Above all, every boater should know when it is safe to venture out and when it is prudent to remain in safe harbor.

Why bother, you may think, when forecasts are readily available on radio and TV, from newspapers and marine weather channels. Unfortunately, though generally helpful, these sources are not always specific enough. Commercial radio and television forecasts concentrate on what landlubbers want to know: Do I need a coat this morning? Will I get soaked at the ball game? Newspapers print weather maps that cover the entire country, but the information they impart is cursory at best. Even marine weather reports are not always reliable. As we all know, weather can change abruptly, and the marine reports are only updated every few hours. Besides, they too report average conditions over a wide area. In Southern California, for instance, the marine report covers "outer waters from Point Conception to the Mexican border." That's about 200 miles in a straight southeasterly line!

Obviously, when predictions call for high winds, it is foolhardy to venture forth. But what if conditions are fair?

Fig. I-1 Weather determines when to leave safe harbor. (Photo by author)

Will they remain that way? Before you leave the marina, can you tell whether you will be caught in a blow or heavy fog? Once out there, can you determine when you should run for shelter? And what about the offshore voyagers who find themselves on the open ocean with the nearest harbor a thousand miles distant? Even then, by being alert to warnings in the sky and on the water, a sailor can usually take actions to mitigate the effects of an impending severe storm.

Meteorology is not an exact science—even the professionals are sometimes wrong in their forecasts. As a boater, you can't always be certain, but you can learn to read the signs and develop a high degree of weather awareness. Then, if you err, you can do so on the side of safety and caution. Above all, you can prepare yourself, your crew, and your boat for whatever the weather may have in store. This preparation will force you to consider what types of conditions create a hazard for your particular vessel and how you will handle precarious situations if you cannot escape them. After all, it is when the unexpected strikes that we find ourselves in trouble—with too much sail up when no one wants to risk the foredeck, when everything not stowed below flies about the cabin, when suddenly the shelter you seek lies dead to windward, or when you find yourself enveloped in thick fog without bearings to guide you. One need only listen to the panic calls on channel 16 when severe weather overtakes the Sunday fleet to know the problems that can be encountered by those who are unprepared.

Two of the primary joys of boating are self-sufficiency and intimacy with nature. Knowledge of weather conditions increases boating pleasure on both these counts. When you know what the barometer tells you, when you know what it portends when the wind backs or veers, when the seas increase to windward, or when clouds thicken and lower, you can develop an intimacy with the ever-changing forces in our environment and a confidence in your ability to handle your boat safely.

The story of weather is a fascinating one. There are definite forces that determine the basic weather patterns in different parts of the world, but within those patterns are variations due to land configurations and distant storms that must be taken into account.

The first part of this book is devoted to an in-depth explanation of the forces that shape our weather at different times of the year in different locations. The second part deals specifically with the immediate relationship of weather and the boater. Those readers who are interested only in the practical applications of the subject can start with the second section and refer back to the first part to answer any questions that may arise. Either way, this book is designed to raise the weather consciousness of boaters so that all may enjoy their sport with greater pleasure and safety.

Part I

Forces That Shape Our Weather

1. Climate and Weather

Day became night. Towering black clouds ripped open, disgorging torrents of water—water whipped into horizontal frenzy by screaming winds. The tempest lashed the tops off waves, which grew increasingly mountainous as the storm raged on. Long streaks of biting spray were everywhere, illuminated by relentless bursts of lightning. Giant rollers rushed ahead as the hurricane surged across the warm ocean waters to the accompaniment of howling winds, roaring seas, pounding rain, and crashing thunder, threatening death and destruction for anything or anyone within the storm's path.

Although most of us will never personally experience a hurricane, we can often feel its effects thousands of miles away. Even after it has spent its fury, a major storm can spawn tornadoes, waterspouts, high surf, and unseasonable weather along coastlines and even inland in areas not touched by the storm itself.

When we think of weather in a particular area, we must remember that it is affected by weather in other parts of the world. Our planet is one interdependent unit because the atmosphere surrounding it is in constant motion, moving from place to place and picking up the characteristics of heat and humidity of the terrains over which it passes. The route it has followed has a significant effect on the kind of weather that can be expected. In order to understand this, it is necessary to examine our planet as a whole and see what creates our seasons and our weather.

When we speak of *climate*, we are talking about the seasons—about averages of temperature ranges, humidity, rainfall, and winds that dominate at different times of the year in different areas. The word comes from the Greek *klimat*, meaning "inclination" or "tilt." Although climatic conditions change over the years, climate is more or less predictable at any given period of time. *Weather*, on the other hand, refers to the variations that occur on a day-to-day basis within the overall climatic parameters. It is the state of the atmosphere at a definite time and place with respect to temperature, precipitation, and wind. For instance, during a rainy season, some beautiful sunny days will occur.

THE EARTH'S JOURNEY AROUND THE SUN

Because the earth revolves around the sun at a tilt of 23.5°, (fig. 1-1), the angle of the sun's rays reaching the earth's surface changes during its annual journey of 365.25 days. Astronomers and navigators talk about the "geographical position" of the sun when referring to that place on earth (latitude and longitude) where the sun is directly overhead at a particular moment. The sun's position with respect to the earth's latitude is known as its *declination*.

Four days in every year have special climatic significance. All are directly related to the 23.5° tilt of the earth. On or about June 21, on the summer solstice, the sun has reached its most northerly point and is directly over the parallel of latitude 23.5°N, which we call the Tropic of Cancer. On that day, the Northern Hemisphere begins its summer: The sun's rays are more vertical and intense and warm the Northern Hemisphere for more than half of each day. The corresponding date in the Southern Hemisphere is on or about December 22, when the sun is directly over

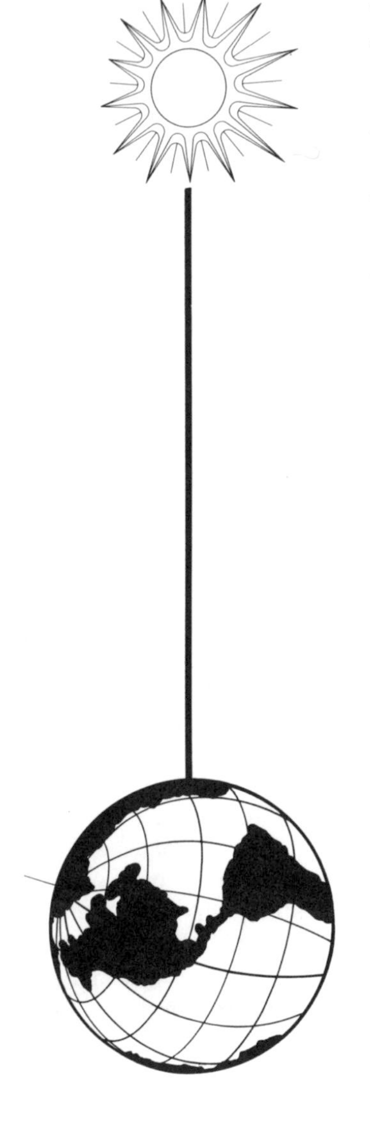

Fig. 1-1 The relation of the earth to the sun at the summer solstice. The earth is always tilted at a 23.5° angle.

its most southerly latitude—23.5°S, the Tropic of Capricorn. On this date, the sun's rays reaching the Northern Hemisphere do so at their most extreme oblique angle, producing the shortest day and longest night of the year. It is winter.

The word *solstice* means "sun stands still" because the sun tarries a while on that latitude before it appears to change directions. The term, of course, originated in a time when it was thought that the sun moved around the earth; people could see that sometimes it was moving in a northerly direction and at other times in a southerly direction. Between the latitudes of 23.5°N and 23.5°S, at some time during the year, the sun is directly overhead, and its rays reaching the earth at that time are vertical. This area is known as the *tropics.* No other area of the earth is ever reached by vertical rays of the sun.

Twice a year, the sun is directly over the equator at a declination of 0°. These are the times of the equinoxes. During the vernal equinox, on or about March 21, the declination of the sun shifts to the Northern Hemisphere, and on or about September 23, the autumnal equinox, its declination shifts to the Southern Hemisphere. On those days, both hemispheres receive the same amount of sunshine and the days and nights are of equal duration—hence the term *equinox* ("equal nights") (fig. 1-2).

To complete the picture, we must consider the polar regions. During the respective summer months, from the poles to a latitude of 66.5° (23.5° from the poles), these areas enjoy continuous sunlight, but during the winter months, they are cloaked in perpetual darkness. Those parallels of latitude—the Arctic Circle and the Antarctic Circle—describe the limits of what is known as the circumpolar sun.

Length of Daylight

North Latitude (Degrees)	*Vernal Equinox March 21*	*Summer Solstice June 21*	*Autumnal Equinox Sept. 21*	*Winter Solstice Dec. 21*
0	12h 07m	12h 08m	12h 07m	12h 07m
30	12h 09m	14h 04m	12h 07m	10h 13m
60	12h 16m	18h 52m	12h 12m	5h 53m
90	—	24h 00m	—	00h 00m

Fig. 1-2 Length of daylight in the Northern Hemisphere, calculated from times of sunrise and sunset given in the Nautical Almanac, U.S. Naval Observatory.

THE EARTH'S ROTATION

The earth completes an eastward rotation on its axis every twenty-four hours. This means that at some moment during each day, the sun's geographical position is directly over every longitudinal location on the earth. The geographical position of the sun with respect to longitude is known as the GHA, or *Greenwich Hour Angle.* This angle changes by 1° of longitude on earth every four minutes—resulting in day and night and the consequent heating and cooling that occur during different times of the day.

The effect of this rotation on the atmosphere goes beyond that. When we look at a globe, it is easy to see that the lengths of parallels of latitude decrease poleward. As the earth turns, the rate of spin is greatest at the equator and decreases toward the poles. At the equator, the earth spins at a speed of 868 knots. At latitude 60°, its spin is only 434 knots. At the poles, there is no spin at all. This affects the movement of the atmosphere and is known as the Coriolis effect.

THE CORIOLIS EFFECT

If the earth did not rotate, a rocket fired on a north-south trajectory would simply follow a north-south path. However, due to the earth's eastward rotation, its path will be deflected. In the Northern Hemisphere, the deflection is to the right; in the Southern Hemisphere, to the left. The same thing happens to wind. It is deflected and follows a curve rather than a straight line. For instance, winds moving north (in the Northern Hemisphere) will move in a northeasterly direction. Winds moving south will move in a southwesterly direction. The opposite is true in the Southern Hemisphere. The effect is strongest at the poles and negligible at the equator, and it increases with wind speed. (See Chapter 7 for a further explanation of this effect.)

EFFECTS OF LAND

If the earth had a uniform surface, the circulation of air would be simple. But the earth has a varied surface that distorts this circulation as it passes over water, ice, forests, mountains, plains, valleys, deserts, and jagged coastlines. Particularly in the higher latitudes of the Northern Hemisphere, wind patterns over the oceans are affected by the large landmasses, whereas in the Southern Hemisphere, only the tip of South America, a part of New Zealand, and the small island of Tasmania jut into the ocean south of about the 38th parallel, giving the winds of that area a greater constancy.

Climate and weather are complex phenomena because they are subject to many interrelated forces. An examination of each of these forces will show how they are related to the big picture of climate as well as to the smaller weather picture of specific areas.

2. Highs and Lows

The atmosphere, which surrounds the earth in layers like blankets, is an integral part of our planet, bonded to it, like everything else, by the force of gravity. It controls both temperature and climate. Near the surface, it consists of a mixture of gases, primarily molecular nitrogen (78.1 percent by volume) and oxygen (21 percent by volume), interspersed with a variety of other gases, water vapor, and an assortment of solid and liquid particles. When the atmosphere is in motion relative to the earth's surface, we have wind. For us as boaters, wind is the most important weather phenomenon, so it is imperative to analyze why we have wind and what forces determine its strength and direction.

THE TROPOSPHERE

The portion of the atmosphere that affects us most directly is the layer closest to the earth known as the *troposphere.* It is the realm of clouds and precipitation and it extends to an altitude of approximately 11 miles at the equator but only 4 to 5 miles at the poles. It is a region of decreasing temperature with increasing height. Most of the air is concentrated in this layer: More than half of the atmosphere is within less than 3.5 miles above the planet's surface, and 97 percent is within less than 18 miles above it. We tend to lose perspective when considering heights. Eighteen miles above the earth seems very far away until we realize that at 35 mph it takes only about half an hour to drive that distance.

PROPERTIES OF AIR

Although we are not normally aware of it, air has weight. At sea level, at a temperature of 59°F, a cubic foot of air weighs 1.2 ounces, but the column of air above that cubic foot presses down on the surface of the earth with a force of 14.7 pounds per square inch, or about one ton per square foot. By international agreement, this is defined as "one standard atmosphere." With increased height, the amount of air (and consequently the weight of the column above it) decreases—as is well known to anyone who has ever climbed a mountain, hiked the high country, or visited a high-altitude area.

In addition to pressing downward, air also presses laterally, horizontally to the earth's surface. If this pressure were the same all over the earth, there would be no horizontal movement of the air. But when the pressure varies, and inequalities exist in the mass of the atmosphere, the result is wind.

Besides having weight, air is also compressible, expanding and contracting with changes in temperature. Like other objects, air expands when heated and contracts when cooled. A simple experiment demonstrates this point. If you blow up two balloons to the same size and put one in a refrigerator, the refrigerated one will become much smaller than the one that remained at room temperature. The amount of air within the cooled balloon remains the same, demonstrating that for a given space, more air is contained in a cooled area than in a warm one. Hence, cooled air is heavier, by volume, than warm air. When we use terms such as *cold* and *warm*, we must remember that these are relative, not absolute terms. What feels cold at the equator may feel very warm at the poles. The classic experiment that proves this point involves three bowls of water—one hot, one lukewarm, and one cold. If you place your left hand in the bowl of hot water and your right hand in the bowl of cold water, leave them for a few minutes and then put both hands in the lukewarm water, the left hand will sense that water as cold while the right hand will sense it as hot.

As air is heated, it expands, resulting in less air by volume and consequently less pressure both downward and outward. As air cools, the reverse takes place: It becomes heavier and pressure increases both downward and outward. The result is that the cooled air pushes against the less dense warm air and moves it out of the way. For this reason, air tends to move from an area of high pressure to one of lower pressure.

As the warm air rises, it flows toward cooler areas to fill the void left when the cooler air compressed. It, in turn, is cooled at the higher elevations, becomes heavier, and sinks back to earth, thus completing the cycle.

Air is also absorbent. It takes up salt particles from the oceans and sand, dust, and pollutants from the land over which it passes. And it absorbs moisture. The higher the temperature, the more moisture it can absorb. For this reason, low pressure frequently is associated with high humidity, clouds, and precipitation, whereas high pressure usually means dry weather with clear skies.

DIRECTION OF WIND FLOW

Wind direction is affected basically by three forces. One of these is the Coriolis effect (see chapter 1), which deflects the wind to the right of its path (in the Northern Hemisphere).

Another is the frictional force. As wind travels over land, it encounters friction, which acts in the opposite direction to the wind, slowing its progress. Since land creates more friction than water, surface wind can attain far greater sustained velocities over large bodies of water than it can over land. At greater elevations, friction becomes less and less of a factor—until, at about 3,000 feet above the surface, it becomes negligible. Thus, wind velocity increases with height above the earth. From wind measurements we learn that winds at 3,000 feet are 2.5 times stronger than over land and 1.43 times stronger than over water.

The third factor determining wind direction is the pressure force. It acts opposite to the combined Coriolis and frictional forces, causing the wind to blow at an angle from a high-pressure area to one of low pressure, rather than in a straight line.

When we consider the above forces, it becomes evident that where wind flow is concerned, nothing happens in a straight line. Directions are always curved, circular, or elliptical. In an area of low pressure, winds from a high-pressure area will flow around the low in a counterclockwise direction (in the Northern Hemisphere), trying to displace the low-pressure system. Conversely, the high, abutting this low, deflects the wind flow over and around it in a clockwise direction, much as gears meshing together must turn in opposite directions. The counterclockwise flow is termed *cyclonic* and the clockwise flow, *anticyclonic*. In the Southern Hemisphere, the direction is reversed, with the flow becoming clockwise around the low and counterclockwise around the high.

ISOBARS AND WEATHER MAPS

By means of a variety of weather instruments, meteorologists are able to accurately measure atmospheric pressure over the earth. On weather maps, lines are drawn to join areas of equal pressure. These lines are known as *isobars*. Isobars are always curved; many are elliptical and show the high- and low-pressure areas that exist at any given time over the globe. These areas may vary from a few hundred miles in diameter to about two thousand miles.

A weather map closely resembles a topographical map. On the latter, the closer the altitude demarcations, the steeper the terrain. On a weather map, also called a *synoptic chart* (because it provides a synopsis of meteorological conditions), the closer the pressure lines, the greater the wind strength. In addition, the distribution of highs and lows on the chart enables predictions to be made as to the direction of the wind, since it will flow from high-pressure areas to those of lower pressure (fig. 2-1).

In addition to the charting of surface pressures, upper-level synoptic charts are also prepared. These show the direction of the upper-level steering currents that determine the general flow of the surface weather. Besides the locations and shapes of highs and lows, other pertinent information is available on weather maps, which will be discussed in a later chapter.

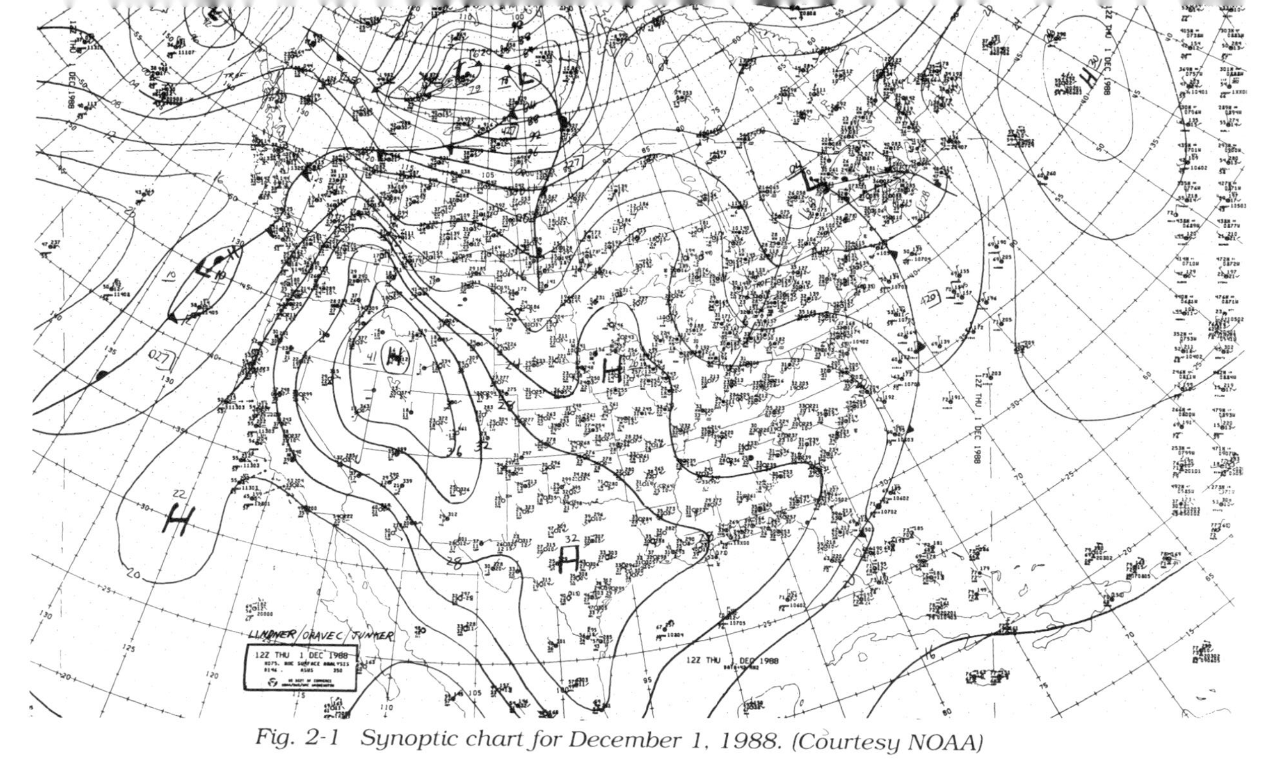

Fig. 2-1 Synoptic chart for December 1, 1988. (Courtesy NOAA)

3. Heating and Cooling

Of all the interrelated forces that produce weather, heat is the most powerful—so powerful that it frequently is referred to as the engine that drives the general atmospheric circulation.

HEAT AND TEMPERATURE

Although often used interchangeably, the terms *heat* and *temperature* have different meanings. Heat, measured in calories or British thermal units (Btu), is a form of energy caused by the ceaseless motion of tiny, invisible atoms and molecules. These microscopic particles may move slowly or at frantic rates. The faster they move, the hotter the object and consequently the greater the heat imparted to the surroundings.

Temperature, on the other hand, is a measure, in degrees of Celsius or Fahrenheit, of how hot an object is. For instance, a cup of boiling water has the same temperature as a pot of boiling water, but the latter contains a great deal more heat than the former. The addition of heat is required to raise the temperature of an object and the subtraction of heat is necessary to lower the temperature of that object.

SOLAR RADIATION

The sun's heat energy is emitted in the form of waves, which have crests and troughs much like ocean waves. The greater the heat, the shorter the distance between the crests. Since the sun's heat is so intense, solar radiation is in the form of short waves, and it includes, among others, those wavelengths we perceive as light.

Of the heat energy radiated by the sun in all directions into space, less than one-billionth actually reaches the earth. That small fraction is so intense that it exceeds by several hundred thousand times the total electrical generating capacity of the United States. Of the portion that does reach the earth, some 18 percent is absorbed by the oxygen and ozone in the atmosphere; 35 percent is reflected back into space by clouds, the earth, and the atmosphere; and only the remaining 47 percent or so is actually absorbed by the earth's surface.

Clouds, dust, and other particles in the atmosphere tend to reflect, scatter, and absorb some of the radiation falling on them. As a result, solar radiation reaching the earth in areas where dense clouds are the norm is far less than in places where clear skies permit the solar rays to penetrate to the surface of the earth.

Even though the sun's rays always have the same intensity, the amount of heat reaching the earth varies not only with atmospheric conditions, but also with the time of day and the season. At sunrise and sunset, when the sun's angle is low, its slanting rays are spread over a large area, and the heat received at any one point on the ground is less than at noon, when the sun is high and its rays are concentrated vertically on a smaller area. The same is true of the seasons, when the heat we receive depends on the geographical position of the sun as well as on the latitude of different places in the world (fig. 3-1). For instance, when the sun is over the Northern Hemisphere, the rays reaching that part of the earth are more vertical and intense than those reaching the

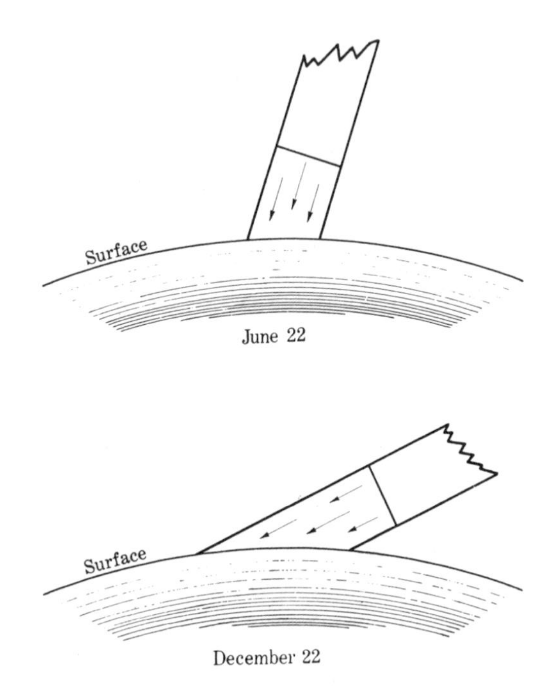

Fig. 3-1 Sunlight in summer and winter. Although the earth's rays always have the same intensity, the more slanting the rays, the greater the area over which the heat is spread.

Southern Hemisphere. Hence it is summer in the former and winter in the latter.

Much of the radiation that does reach the earth is wasted—reflected away at greatly varying rates: 80 to 90 percent from newly fallen snow; 30 to 35 percent from deserts; 5 to 40 percent from forests (depending on the color of the leaves); 8 to 25 percent from crops and grasses; 2 to 5 percent from water surfaces when the sun is high (but nearly 80 percent when the sun is near the horizon).

Nor is the sun the only body that radiates heat. Nearly everything on earth does the same. But since other bodies are much cooler, the energy they radiate is of far greater wavelength. This long-wave radiation is readily absorbed by the water vapor, clouds, and carbon dioxide in the atmosphere, from where it is reradiated to earth. From this sequence of events, we can understand why, under a cloudless sky, the days are hotter and the nights colder than in the presence of clouds—which, on the one hand, keep some of the sun's rays from reaching the earth, but on the other hand conserve the heat and prevent it from escaping into space.

LAND AND WATER

Every location on earth experiences day and night as well as climatic changes of the seasons. But there is a vast difference in weather conditions between an inland area and a coastal one, even if they are at the same latitude. There are several reasons for this. Two of the most important are:

1. Land surfaces are opaque, which means that heat is absorbed only at the surface. As a result, rocks or sand will become hot quickly on a sunny day and cool just as quickly when the sun's heat is replaced by the darkness of night. Water, on the other hand, is transparent, permitting solar radiation to penetrate to a depth of about thirty-three feet. Through motion in the water, this heat is distributed fairly evenly throughout that depth. Water, therefore, requires more heat to be applied over a longer period to raise its temperature. When the heat source is removed, it draws upon the heat reserves stored at depths below the surface. The cool water sinks, permitting warmer water to rise.

2. Every object requires different quantities of heat to raise its temperature by one degree. This is known as the *specific heat* of the object. Water has a specific heat of 1. For most rocks, it is only .2. This means that it takes five times more heat to raise the temperature of one ounce of water by 1°F than it would to raise the temperature of one ounce of rock by 1°F.

For these reasons, land areas respond quickly to changes in radiation, whereas bodies of water heat and cool slowly. To cite two extreme examples:

1. The sand surface of the Sahara Desert changes temperature by more than 70°F from day to night.
2. The water-surface temperature in the tropics varies only 2°F from day to night.

In addition, the constant temperature of bodies of water serves to moderate the temperatures of coastal areas, and, conversely, rapid changes of temperature on land influence weather conditions on the waters along a coast and frequently far out to sea. However, beyond the influences of the land, weather conditions on the oceans tend to be more stable and consistent than they are on or near land.

OCEAN CURRENTS

Clearly defined currents are found in many parts of the oceans (fig. 3-2). The warm Gulf Stream and its extension, the North Atlantic Drift, bring warm water from the tropics to western Europe. This moderates the winter temperatures in a place such as London, where the average temperature during January is 8°F higher than it is in New York, even though London is at a latitude of 51°N and New

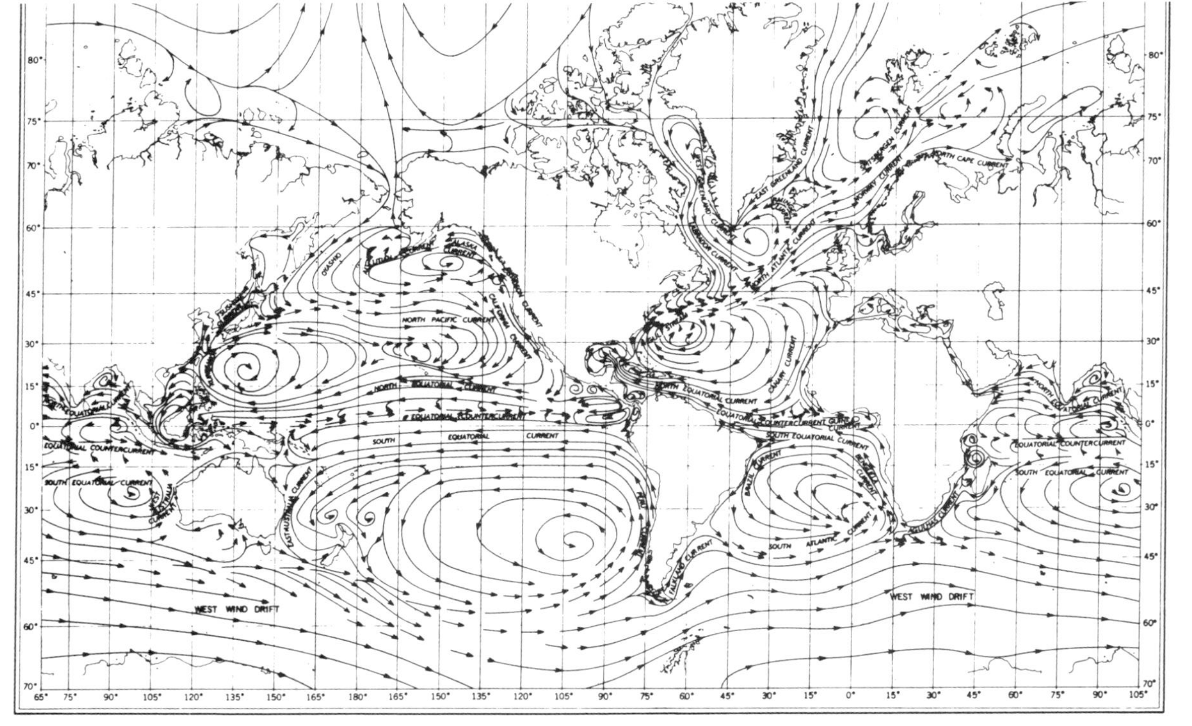

Fig. 3-2 Major surface currents of the Northern Hemisphere. (Courtesy Bowditch)

York is at 40°N. On the other hand, the cold California Current, which flows south from arctic waters, keeps the summer temperatures in Southern California cooler by about 10°F than in the corresponding regions of the East Coast of the United States.

The circulation of the oceans is much slower than that of the atmosphere. Therefore, the warmth that reaches the European coast in the winter originated over low latitudes in the Atlantic some six months earlier.

THE HEAT BUDGET

Over the entire area of the earth's surface, incoming radiation equals outgoing radiation, thus keeping the average temperature of the earth rather constant. This is known as the *heat budget*. However, this balance is not maintained in all regions. At latitudes below 45°, incoming radiation exceeds that lost to space, whereas at higher latitudes, more radiation is lost than is received. This does not mean, however, that the poles are getting colder and the tropics hotter, because the atmosphere (and to a lesser degree the oceans) acts to transfer heat from the tropics poleward. It is this imbalance of heat that drives the winds and the ocean currents.

As we saw in chapter 2, in the upper atmosphere, warm air flows toward regions of cooler air that has compressed and left a void for the warm air to fill. Most of this continuous heat transfer from the tropics to the poles occurs in the midlatitudes, creating the stormy weather often associated with this region.

When visualizing these large-scale exchanges of air from the equator to the poles, it becomes apparent that atmospheric conditions from one area affect other areas as the atmosphere travels over them—proving just how interdependent our planet really is.

4. The Earth's Wind Systems

The earth is divided into a series of surface wind systems, each of which has dominant patterns that prevail, with some variation, throughout the year (fig. 4-1). However, since landmasses distort the patterns, they are truly valid only over large bodies of water and are often not felt for the first fifty or so miles offshore.

THE HORSE LATITUDES

At the equator, hot, moist air is pushed poleward at high altitudes. It cools, becomes more dense, and, when it has traveled about one-third of the distance to the poles, some of it sinks back to earth. When there is such a downward airflow, the upper portion is heated by compression while the temperature of the portion near the surface remains unchanged.

Whenever temperatures equalize, the result is a stabilization of the air. This evaporates the clouds in the atmosphere and results in clear skies over a region. It also results in weak pressure gradients. On land, this is the region of the world's tropical deserts. At sea, it is characterized by light and variable winds, scorching sun, and little precipitation. The *horse latitudes* were named during the era of sail: Square-riggers frequently were becalmed for days at a time under a scorching sun, and horses and other animals aboard died from heat and lack of water and had to be thrown overboard.

In this region, the airflow splits into a poleward flow and an equatorward flow. The latter forms the trade-wind belt and the former the westerlies of the midlatitudes.

THE TRADE-WIND BELT

As the winds from the high-pressure area of the horse latitudes blow toward the low-pressure area around the equator, they pass over a region known as the *trade-wind belt*. Due to the Coriolis force, the trade winds are northeasterly in the Northern Hemisphere and southeasterly in the Southern Hemisphere. Frequently they blow for days, even weeks, with great constancy and little change in

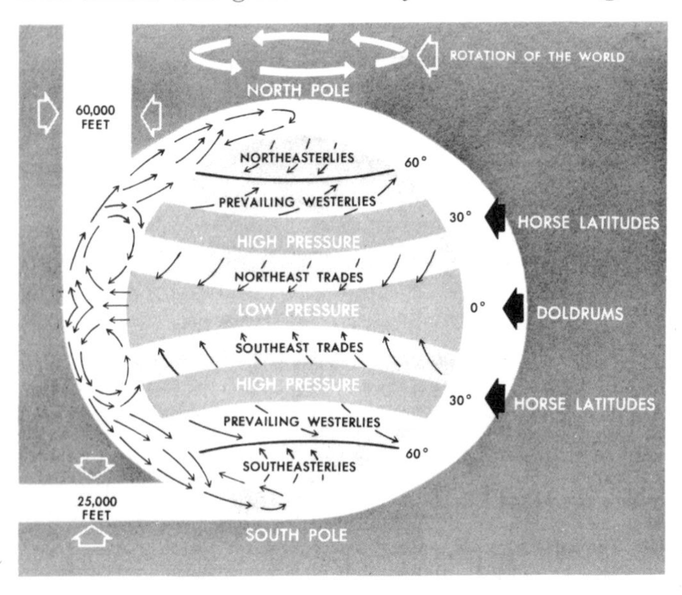

Fig. 4-1 The earth's major wind systems. (Courtesy Bowditch)

direction. In the summer months, 20- to 30-knot winds are the norm, with the intensity abating after dark and gradually increasing again after dawn. They are stronger everywhere during the winter months. Unlike in the horse latitudes, skies here frequently are cloudy (fig. 4-2).

The trade winds are prevailing winds: Their pattern remains generally consistent, and the variations in direction and intensity that may occur are not common and rarely last long. These changes usually are caused by storms that have drifted from the usual tracks.

THE DOLDRUMS

The area near the equator that extends approximately from 10°S to 10°N latitude is known as the *doldrums*. It is also called the *intertropical convergence zone*, because the trade winds from the two hemispheres meet and mingle here.

Fig. 4-2 Running down the trades in steady winds under cloudy skies. (Photo by author)

Due to the intense heat and the consequent high moisture absorbency of the air, this is always an area of low atmospheric pressure. It is generally cloudy, with showers and thundershowers occurring frequently because of rising currents of warm, moist air, which condenses as it cools at higher elevations. Due to the slight daily temperature variations, winds generally are light and fluky, frequently nonexistent. But the area is also a zone of highly unsettled conditions that can change without warning from windless to hold-onto-your-hat. For this reason, some sailors who have ventured through this area call it the "intertropical crazy zone."

THE PREVAILING WESTERLIES

The next section of the poleward flow that originates in the horse latitudes results in the prevailing westerlies of the temperate zones. Again, due to the Coriolis force, these winds are deflected, becoming southwesterly in the Northern Hemisphere and northwesterly in the Southern Hemisphere. In the Northern Hemisphere, even over the oceans, the prevailing pattern is greatly distorted by the large landmasses that lie between 40° and 50°N. In addition, since this is the area of much of the heat exchange in the upper atmosphere, stormy weather and frequent changes in wind strength and direction are common. The wind strength of the prevailing westerlies is considerably higher in winter (when the average is 25 knots) than in summer (when the strength drops to an average of 14 knots). Since the Southern Hemisphere does not have any large landmasses at those latitudes, the westerlies there are stronger and more constant throughout the year—hence their nickname, *the roaring forties*. The farther south you travel, the more fierce is the roar.

THE POLAR REGIONS

Winds in the polar regions are northeasterly in the Northern Hemisphere and southeasterly in the Southern Hemisphere, thus moving toward regions of lower pressure at lower latitudes. In general, the circulation in the arctic region is modified by the surrounding landmasses. In the antarctic, however, a central landmass, surrounded by water, serves to increase the circulation, causing the winds to remain strong throughout the year—sometimes as much as 200 knots at the surface. Except for some hurricanes, these are the strongest winds found anywhere on the earth.

As these cold polar winds move toward the equator, they encounter the warmer, westerly flow of the midlatitudes. The region where these opposing flows clash is known as the *polar front.*

THE WINDS ALOFT

The wind patterns discussed here have been those felt on the surface of the earth. But the hot tropical air that rises aloft sets up its own motion in the upper atmosphere, and, as mentioned previously, this upper airflow can reach far greater velocities than surface winds, since the element of friction has been removed.

There is a limit to how high air can climb in the atmosphere because there is a ceiling past which it cannot penetrate. That ceiling, known as the *tropopause*, constitutes a barrier between the *troposphere* and the *stratosphere*, the next higher layer of atmosphere.

The stratosphere is very different. It is an area of clear skies and slight temperature variation. In its lower section there is a slow, constant increase in temperature with altitude. This temperature increase becomes more rapid with increasing height, reaching a maximum of about 32°F near the top of the stratosphere, some 30 miles above the earth. There is little exchange of air between the troposphere and the stratosphere, indicating that there is a distinct difference in the circulation between the two layers.

Since the average air pressure at sea level is about the same everywhere, it follows that heated, expanded air reaches greater heights near the equator than near the poles. Therefore, the location of the tropopause varies from about 3 to 5 miles near the poles to 8 to 10 miles near the equator.

As a result of aircraft activity during and since World War II, much has been learned with respect to upper-level airflow. One of the findings that is most pertinent to an understanding of global weather is the fact that at most latitudes—except near the equator where the Coriolis force is weak—the airflow in the mid and upper troposphere is westerly: from west to east. This occurs because the predominant upper-level airflow is from the equator to the poles—in other words, from south to north. Add the Coriolis force and a strong westerly component results. This is demonstrated daily on television weather reports, where weather systems are shown moving inland from the Pacific toward the Atlantic, across the United States.

JET STREAMS

During winter, in the midlatitudes, great temperature contrasts frequently exist in areas separated sometimes by as little as a few hundred miles along regions known as *fronts*, where cold and warm air masses meet. This happens over land areas, but also at sea, particularly in areas of strong ocean currents. When the edge of a warm current abuts a cold one, strong temperature gradients develop in the sea surface. Such inequities affect the general upper airflow. Directly above these regions of great temperature contrasts are very fast currents of air known as *jet streams.*

The existence of jet streams was first verified during World War II, when U.S. bombers headed westward to Japanese-occupied islands and found that sometimes they could make little or no headway. When they turned back toward their base, they experienced tail winds that at times exceeded 160 knots. In the midlatitudes, a jet occurs in conjunction with a polar front, which, as mentioned previously, is located between the cold polar easterlies and the warmer westerlies. Other jets exist in the world, but this polar jet has the strongest influence on weather in the United States, so it is the most important for our purposes.

Despite the fact that upper-level air generally flows in a west-to-east direction, the polar jet often wanders widely from this course, at times dipping north, at other times south. When it dips south, an elongated low is produced, allowing colder air to move southward. When it bends toward the north, a ridge of high pressure results, forcing warmer air toward the poles.

The atmosphere is in constant search of equilibrium. Air flows in order to bring pressures and temperatures into balance. We have discussed some of the systems here, and in the following chapters will consider more of these processes.

5. Temperature and Humidity

Another vital component of weather, so necessary for life on this planet, is the never-ending circulation of the water supply from the ocean to the land and back again—a process known as the *hydrologic cycle.*

THE HYDROLOGIC CYCLE

At any particular time, the amount of water vapor (a colorless, odorless gas in the air) is almost insignificant—somewhere between 0 and 4 percent by volume—but it has been estimated that the total amount of water distributed by the atmosphere to different parts of the earth in one year is enough to cover the entire surface of the globe to a depth of about three feet.

Water—mainly from the oceans but also from other sources—is constantly evaporating into the air, from where it is transported by the wind to distant areas. When conditions are right, the water vapor is released by clouds that have formed, and falls back to earth as precipitation. It is estimated that the average time water spends in the air between its evaporation and its return as precipitation is nine days. Some water falls directly into the ocean or into lakes and rivers, from where it is returned to the ocean. The rest falls on land, soaks into the ground, and from there makes its way, via lakes or streams, back to the ocean. As with the heat budget, which maintains a constant average heat over the entire globe, the hydrologic cycle maintains a relatively stable water exchange. Over the oceans, evaporation exceeds precipitation; over land, the converse is true.

PROPERTIES OF WATER

Water is the only component of air that can change its state directly from a gas to a liquid or even to a solid and back again. Such changes of state are set in motion by heat from the sun.

We know that heat is required to melt snow—to change a solid into a liquid. The same is true when a liquid is changed into a gas. Heat energy applied to water increases the motion of the surface molecules enough to allow them to evaporate—to escape into the air. Only the more rapidly moving surface molecules escape into the space above the water. Since the slower molecules remain in liquid form, the temperature of the water drops. In other words, the heated molecules that have vaporized have taken the heat with them and are storing it until they return to liquid, at which point they again release what is called their "latent" heat. We can feel this cooling process after a summer shower, when evaporation from wet ground leaves the air deliciously cool and refreshing. The rate of evaporation from a water surface increases as the temperature rises and the wind strengthens.

The converse occurs during condensation. When the water vapor changes back to a liquid by cooling, the heat that originally was absorbed during evaporation is released. Since evaporation is greatest over tropical oceans and the water vapor is transported by the atmosphere to other parts of the earth, the hydrologic cycle provides another method of circulating heat from warmer to colder regions.

RELATIVE HUMIDITY

Relative humidity is the ratio (expressed as a percent) of the actual quantity of water vapor in the air to the amount of vapor the air can hold at a given temperature. Since warm air is more absorbent than cool air (see chapter 2), the warmer the air, the more moisture it can hold before it reaches the saturation point—the point at which it cannot hold any more. Therefore, when the air contains a certain amount of water vapor by volume, and the temperature rises, the relative humidity falls. If the temperature falls, the relative humidity rises until the cooled air is saturated and condensation ensues in the form of precipitation, dew, fog, or mist.

If the air is laden with other particles—such as dust, salt spray, smoke particles, or pollutants—it may release its moisture before it has reached the saturation point, because these particles are *hydroscopic*—they attract moisture (more about that later).

WIND, TEMPERATURE, AND HUMIDITY

Temperature and humidity are affected by the terrain over which they pass. If the wind is calm and the atmosphere passes slowly over an area, perhaps even lingers a while, it will come into harmony with the surface of the earth. If the wind is strong and the air is moving rapidly, the adjustments will be small. Therefore, in order to understand weather in a particular place at a certain time, it is important to know what path the atmosphere has taken to reach it.

There are four basic paths the atmosphere can take. If it has come from warm oceans, it is called *maritime tropical* (mT) and will be moist and warm; if from cold oceans, *maritime polar* (mP)—moist and cold. If it has come over warm land, it is called *continental tropical* (cT) and will be dry and warm, and if it has come over cold land, it is classified as *continental polar* (cP)—dry and cold.

CLOUDS

Whenever we see a picture of our planet taken from outer space, we see portions of it covered by clouds of myriad shapes and sizes. Clouds form when warm, moist air rises. As this air enters regions of ever-lower atmospheric pressure, its temperature is reduced. As the temperature drops, the air can hold less and less water vapor, until there is more than it can contain. The excess then condenses into tiny droplets, which, taken together, become visible as clouds. Each of these droplets is only about one-eighth the diameter of a human hair, and consequently so light in weight that it falls very slowly. An average droplet, falling from a cloud at 3,000 feet, would journey about forty-eight hours to reach the ground. Obviously, such droplets evaporate at lower, warmer altitudes and do not produce precipitation. Rather, they remain in the atmosphere as moisture (fig. 5-1).

FOG

As we have seen, when water evaporates, the vapor hovers over the surface of the water. If there is upward motion in the air, the vapor will rise to form clouds at higher altitudes. If the motion is lateral, the warm, moist air passes over the cooler water and the vapor forms a cloud just above (perhaps touching) the surface of the water. That cloud, if dense and extensive, prevents us from seeing one

Fig. 5-1 Clouds form when warm, moist air rises. (Courtesy NOAA)

end of the boat from the other and is that often-frightening phenomenon called *fog*.

PRECIPITATION

In order for clouds to produce precipitation, other processes must take place. As far as is known, there are two distinct ways in which this can occur. The first of these takes place mainly in tropical and subtropical areas; the other occurs in higher latitudes.

As warm, moist air rises, it is cooled and its relative humidity increases until the air becomes saturated. As it cools further, it becomes supersaturated and releases droplets of liquid. These droplets are not all the same size because they form around hydroscopic nuclei of varying sizes present in the atmosphere. The bigger ones will fall faster than the smaller ones. As they fall, they bump into other droplets with which they combine (coalesce). Particularly in deep clouds, when currents of moist, rising air are strong, the larger droplets spend enough time bumping into the smaller ones, coalescing with millions of them, to attain raindrop size.

At higher latitudes, the formation of precipitation is generally dependent on the presence of ice crystals. In the rarefied atmosphere of higher altitudes, water does not freeze at 32°F. In fact, water suspended in air high in the troposphere may not freeze until its temperature is lowered to –40°F.

To understand this, consider the boiling point of water. When water in a glass container is heated, bubbles can be seen forming near the bottom. As the heat increases, the bubbles begin to rise to the surface, growing in size as they do so. That temperature at which the bubbles begin to reach the surface of the liquid is called the *boiling point.* But atmospheric pressure is pushing down on the water, trying to keep the bubbles submerged in the liquid. Near sea level, the temperature at which the bubbles break the surface of the water is 212°F. If the pressure is reduced at higher elevations, the bubbles have an easier time on their way up, so the molecules don't have to wiggle quite as fast. In other words, less heat is required for the water to come to a boil because there is less pressure to hold the bubbles in the liquid. Anyone who has ever tried to cook potatoes high in the mountains knows that it takes much longer than it would at sea level. Even though the water is boiling in both locations, it contains less heat at the higher elevations. The converse is true in the case of a pressure cooker, where the increased pressure demands more heat to be developed before the water will boil, thereby reducing the cooking time for the food being prepared.

From the above, it follows that if the boiling point of water is lowered under reduced pressure, the freezing point also will be lowered. Thus, water can be cooled well below the normal freezing point and still remain in a liquid state. Water that remains liquid at temperatures below 32°F is termed *supercooled.* Supercooled droplets will freeze if disturbed or if they come into contact with solid nuclei that have a crystal form resembling ice—*ice crystals.* These are sparse in the atmosphere but, when surrounded by great numbers of supercooled droplets, the mixture becomes unstable.

In addition to the crystals and droplets, water molecules also are present in the upper atmosphere, which usually is only slightly supersaturated with water vapor (molecules) with respect to the droplets but highly oversaturated with respect to the ice crystals. To even things out, the crystals begin to gobble up the excess vapor and grow larger and larger. As the ice crystals remove the vapor, the water droplets evaporate, turning into vapor and providing ever more moisture for the ice crystals to absorb. When water freezes, it expands because larger spaces are left between the molecules than were present when the water was in a liquid state. This is what happens when water freezes in pipes, causing them to burst. So the crystals grow quickly and soon become large enough and heavy enough to fall. As they do so, they intercept cloud droplets that freeze on contact with them. A chain reaction develops, producing more and more crystals, which form into larger ones called snowflakes. If the surface temperature is cold enough, precipitation is in the form of snow. If the surface temperature is above 40°F, the snowflakes fall in the form of rain. Even a summer rain may have begun as a snowstorm in the upper troposphere.

6. Types of Clouds

Clouds are a visible clue to what is going on in the atmosphere at any given time. Learning what that is can be difficult, because frequently several types combine to confuse the novice. But with practice, it is possible to predict what type of weather is likely to occur.

Clouds are classified in accordance with their appearance and their altitude. Since altitude is difficult to determine, the shape of the clouds can also give an indication of their height.

CLOUD IDENTIFICATION BY APPEARANCE

There are three basic cloud forms:

1. *Cirrus*, meaning "curl" or "tuft of hair," are the thin, threadlike, delicate ice-crystal clouds that often form long streamers with hooks or curls on the end. Sometimes they are called "mares' tails."
2. *Cumulus*, which means "heap" or "mass," are the dome-shaped woolpack, billowy clouds often seen in summer. They are frequently characterized by flat bases.
3. *Stratus*, meaning "spreading out," are flat, white clouds near the surface. They are in the form of sheets of high fog that cover much of the sky, with no breaks into individual cloud units.

All other clouds are combinations or variations of these three forms.

CLOUD CLASSIFICATION BY HEIGHT

Clouds are also classified as high, middle, or low, depending on the altitudes at which they normally form. Altitudes given here are approximate for the midlatitudes. Their corresponding heights would be higher in tropical regions and lower in polar regions, since the tropopause is lower near the poles and higher near the equator.

High clouds have bases at a height of 18,000 feet or above; middle clouds form somewhere between 6,000 and 18,000 feet; and low clouds appear below 6,000 feet. Because of the low temperature and moisture content at high altitudes, all high clouds are thin and white and composed of ice crystals. Since air at lower altitudes contains more moisture, these clouds are darker and thicker.

Cirrus clouds are composed of ice crystals. Therefore, they are high-altitude clouds. They frequently are brightly colored at sunrise and sunset because, due to their height, they become the first to be illuminated in the morning and the last in the evening. Unless lower clouds also are present, cirrus usually portend fair weather. The appearance of cirrus clouds is also an indication of wind strength aloft. The more brushlike in appearance, the stronger the wind aloft.

Cirrocumulus and *cirrostratus* are variations of cirrus clouds. The former are composed of small, white scales resembling ripples of sand on the beach. They too are composed of ice crystals, and even though they generally are associated with fair weather, they may signal the coming of a storm, particularly if they begin to thicken and lower. In the tropics, they frequently are considered a warning of an approaching hurricane, particularly if they have a reddish-bluish hue when the sun is low.

When cirrus clouds thicken, they become cirrostratus. Just like other stratus clouds, they tend to cover the whole sky, giving it a milky appearance (fig. 6-1). If they continue to thicken and lower, the ice crystals in them will melt to form water droplets, after which they turn into altostratus. When this occurs, rain may be expected within twenty-four hours.

Altostratus (*alto* meaning "high") are middle clouds that also cover much of the sky and have a bluish, veil-like appearance. If these clouds lower and thicken, continuous rain or snow will most likely follow within a few hours.

Altocumulus (fig. 6-2), which also are middle clouds, consist of a layer of ball-like masses, more or less regularly arranged, that frequently merge. They sometimes are confused with cirrocumulus but can be distinguished from these because they are larger and show distinct shadows

Fig. 6-1 Cirrus and cirrostratus. (Courtesy NOAA)

Fig. 6-2 Altocumulus. (Courtesy NOAA)

in some places, which the cirrocumulus do not. When altocumulus thicken and lower, thundery weather and showers may result, although usually only for short periods.

Stratus clouds are low-altitude clouds with a base often as low as 1,000 feet. They are frequently dense and dark, allowing little sunlight penetration. Light mist may result (fig. 6-3).

Stratocumulus are also low clouds. They are shaped like soft, roll-like masses, gray in color, which tend to merge. They move with the surface wind and generally dissipate after dark (fig. 6-4).

Nimbostratus (*nimbus* meaning "rain"), also low clouds, form a dark, shapeless cloud layer, sometimes with ragged

Fig. 6-3 Stratus. (Courtesy NOAA)

Fig. 6-4 Stratocumulus. (Courtesy NOAA)

wet-looking bases. These are rain clouds and can be expected to produce steady or intermittent rainfall but not showery weather (fig. 6-5).

High and middle clouds can lower, thicken, and turn into other types of clouds. There are also some that change in character through upward development. These include the familiar cumulus and cumulonimbus.

Cumulus clouds are billowy, dome-shaped masses that appear in patches, never covering the entire sky. They are marked by strong contrasts between light and dark (fig. 6-6). When there is little vertical development, cumulus are fair-weather clouds, but since they are formed by updrafts, they cause the turbulence that can give airplane passengers a bumpy ride. When the updrafts are strong, cumulus may grow in size and height. When they reach high enough to merge with altocumulus, or develop into cumulonimbus, thundershowers follow.

Fig. 6-5 Nimbostratus. (Courtesy NOAA)

Cumulonimbus are the end result of great vertical development of cumulus clouds. They can rise to considerable heights, and the upper part, consisting of ice crystals, often spreads out into the shape of an anvil. Cumulonimbus are fully developed thunderclouds, producing rain, snow, or hail, frequently accompanied by lightning and thunder (fig. 6-7).

READING THE CLOUDS

By observing the clouds, it is possible to predict whether the day will be wet or dry. It is not possible to tell whether and how hard the wind will blow. Only when considered along with changes in atmospheric pressure is it possible to estimate probable wind strength and direction.

Fig. 6-6 Cumulus. (Courtesy NOAA)

Fig. 6-7 Cumulonimbus. (Courtesy NOAA)

7. Cyclones and Anticyclones

We must now return to some of the concepts covered in earlier chapters—namely, the Coriolis effect (chapter 1) and the lines of equal pressure known as isobars (chapter 2). We have discussed the fact that changes in atmospheric pressure create wind flow and that this flow accelerates from areas of higher pressure to those of lower pressure. But other forces also come into play that affect the behavior of the winds. Although the Coriolis and pressure forces are dominant, we must also consider gravity, friction, and centrifugal force (the tendency of a moving object to continue moving in a straight line).

One factor mitigating wind speed is *friction*, which slows surface winds and changes their direction—but only near the surface of the earth, where the winds actually touch land or water. A few thousand feet above the surface, it is no longer a factor, and it can be disregarded for the following discussion of winds above 3,000 feet.

THE PRESSURE FORCE

Unequal atmospheric pressure drives the winds, moving them from high-pressure areas to those of lower pressure. It moves them perpendicular to the isobars toward the center of the low. The greater the difference in pressure, the greater the wind speed. However, horizontal pressure gradients, like those near the surface of the earth, are small, whereas vertical pressure gradients, variations in pressure with altitude, are great. So why are there not constant strong updrafts in the atmosphere? The answer to this is *gravity*, which balances the vertical pressure gradient and acts in the opposing direction.

In actual fact, wind unaffected by friction does not blow perpendicular to the isobars. Instead, it flows in circular or elliptical sweeps around centers of lows and over centers of highs. It is deflected by the Coriolis force.

THE CORIOLIS FORCE

The Coriolis force balances the pressure force to moderate wind speed and direct its flow. It is not a force in the true sense of the word, but rather an effect—a result caused by the rotation of the earth. For instance, an airplane flying a direct southerly course from a point near the North Pole to the equator will not arrive at its southerly destination at the same longitude because while it is in the air, the earth is turning. In other words, its course is, in effect, deflected to the west. In order to land at the same longitude on the equator, the pilot must adjust his course eastward to compensate.

Similarly, a plane flying eastward from a particular latitude will arrive south of that latitude unless the course is adjusted northward because of the change in angle of the surface of the earth. This angle changes more rapidly near the poles, where longitude lines are close together and thus closer to the axis of rotation.

Substituting wind for the airplane, we find that a wind blowing out of the north actually becomes northwesterly relative to the ground. A westerly wind also becomes northwesterly, since wind is always deflected to the right of its course in the Northern Hemisphere. In the Southern Hemisphere, the direction of the deflection is reversed.

There, both a north wind and a west wind become northeasterly because the relation of the pole to the equatorial bulge has been reversed.

Since the Coriolis force is an effect, it cannot act on still air but increases in strength proportional to the wind strength. To understand this, assume that you are driving along a highway in a strong wind blowing perpendicular to your line of travel. You will feel your car being pushed sideways. In order to stay on the road, you must adjust your steering into the wind. But if you slow your speed, the force of the wind pushing against your car will be lessened depending on how far your speed has been reduced.

If the air were still, only the pressure force could act on it, directing the flow from high-pressure regions to lower-pressure regions. This flow would be across the isobars, which usually are in the form of wide, sweeping curves. As the speed of the flow increases due to the pressure gradient, the Coriolis force comes into play, deflecting the wind. Since the Coriolis force increases with the pressure force, it follows that as wind speed increases, so does the deflection. The Coriolis force is pushing against the pressure force, moving it from its course and back toward the higher-pressure region. Eventually, these two forces come into balance, deflecting the wind to the point at which it flows parallel to the isobars and causing the wind to continue on this course at a constant speed. Because of this, wind strength cannot increase indefinitely—which is happy news for all boaters.

To summarize, the Coriolis force always acts at right angles to the direction of the airflow. It affects only wind direction, not speed: it is strongest at the poles and virtually nonexistent at the equator, so it varies with latitude. Moreover, its effect is directly proportional to the strength of the wind.

RIDGES AND TROUGHS

Sometimes isobars connect to form elliptic or circular cells of either high or low pressure. In the case of a low, the pressure force is directed inward, whereas the Coriolis force is directed outward. In the Northern Hemisphere, where the Coriolis force deflects winds to the right, the wind around a low is coerced into a counterclockwise direction. The reverse is true around a high, causing the wind flow to become clockwise. Circulation around a low is termed cyclonic and around a high, anticyclonic.

When the isobars are curved to form elongated areas of high or low pressure, these areas are called *ridges* or *troughs*, respectively. The greater the curvature, the greater the adjustment the wind must make to follow this curvature. In other words, forces must be strong enough to counter the centrifugal force that acts against this change in direction. Around a high-pressure center, where the outward-directed pressure force is opposed by the inward-directed Coriolis force, the Coriolis force must be strong enough not only to balance the pressure force, but also to turn it inward along the curvature of the high. In the case of a low, the inward-directed pressure force must provide the acceleration to balance the Coriolis force as well as to turn the air.

In the case of circulation around a high, the Coriolis force does the bending. Since it picks up its strength from the wind speed, higher wind speeds are required in an anticyclonic flow than in a cyclonic flow of equal pressure gradients. This does not seem to be what we experience, because we associate strong winds with lows, but the explanation is simple. In the above example, we assumed equal pressure gradients in both cases, but, generally, pressure gradients are greater in a low than in a high. Therefore, anticyclonic winds will frequently be stronger given the same pressure gradients.

THE FRICTION FACTOR

Near the earth's surface, friction between air and land, or water, acts to slow the movement of air and also to change its direction. When the airflow is slowed, so is the Coriolis force. But the pressure force is not influenced by wind speed, so it wins the battle between those two forces and directs the airflow at an angle across the isobars toward areas of lower pressure. That angle depends on the character of the terrain. If it is relatively smooth, such as large bodies of water, the angle becomes roughly 10° to 20° to the isobars and speed decreases by about one-third. Over rough terrain, the angle may become as great as 45° and the speed may be slowed by as much as one-half.

From this discussion of wind direction and deflection, it follows that when an observer on the ground stands back-to-the-wind, low pressure will be to his or her left and slightly in front and high pressure will be to his or her right and somewhat behind. In the Southern Hemisphere, these directions are reversed. This relationship, formulated in 1857 by Buys Ballot, a Dutch meteorologist, is known as *Buys Ballot's law.* The importance of this phenomenon to boaters wishing to predict weather changes is discussed further in chapter 15.

VERTICAL MOTION OF AIR

A question may arise in relation to the fact that the pressure and Coriolis forces seek equilibrium, with the result that wind direction parallel to the isobars continues at a constant speed. What happens to disturb this equilibrium and cause changes in atmospheric conditions? The answer is in the vertical transport of air aloft, caused by conditions in the lower atmosphere.

When air in the lower atmosphere spirals inward toward the center of a low, the inward transport causes the area occupied by the air mass to shrink. This is known as *horizontal convergence.* As the air mass shrinks, it forces the displaced air to pile up, creating a taller and therefore heavier column of air over the low. But the rising air does not stay in one place above the low. It spreads out (diverges) at high altitudes. The surface low can be maintained only if the divergence aloft occurs at a rate equal to the surface convergence. If the convergence exceeds the divergence, the increased pressure of the air column dissipates the low. If the divergence exceeds the convergence, then the upper atmosphere reinforces and intensifies the surface low. Since rising air frequently results in the formation of clouds and precipitation, lows generally are associated with wet weather; if isobars move closer to each other because of the shrinking of the surface air mass, the conditions may be stormy as well.

Low-level convergence can also be caused by other factors. For instance, when air that has been moving over smooth ocean waters hits land, its speed is slowed by the increased friction. It then piles up, causing the air to rise, as in the case of a low. The rising air, as over the low, causes the cloudy conditions frequently found in coastal areas.

Highs have their divergence near the surface, since heavier air columns press outward, leaving space for the atmosphere aloft to converge to fill the void left by the sinking column. As the air journeys earthward, it is compressed and warmed, dissipating the clouds. Highs are therefore associated with fair weather and clear skies, even though they may give rise to strong winds, depending on the proximity and pressure gradient of the nearest area of lower pressure.

From the above, it becomes apparent that weather conditions near the surface are greatly influenced by conditions aloft, which determine the general movement of the atmosphere and serve to intensify or change the weather near the surface.

8. Air Masses

Even though the atmosphere is in constant motion, weather in any particular place is not in a state of constant change. Instead, a weather pattern may remain over an area for several days or even weeks, followed by a shorter period of changing conditions, after which a pattern develops that again remains for a longer period of time. The reason for this is that the atmosphere travels in huge, homogeneous units known as air masses. They are homogeneous mainly with respect to temperature and humidity at any particular altitude.

As discussed in chapter 5, air masses are named for the area over which they develop—a region of the earth where the terrain is basically uniform and weather fairly constant, such as tropical and polar regions (fig. 8-1). Air masses do not form in the midlatitudes because that area experiences great temperature variations and the terrain changes rapidly within the area.

An air mass, moving as a unit, imparts its characteristics to the region of the earth over which it passes. It also mitigates its own nature by picking up other qualities from the earth's surface. The slower the mass moves, the more chance it has to absorb moisture, heat, or cold from the terrain. The faster it moves, the less significant the modification.

An air mass can also be altered when it is above a strong low where rising air currents and divergence aloft dominate. On the other hand, a high near the surface will stabilize the air mass passing over it.

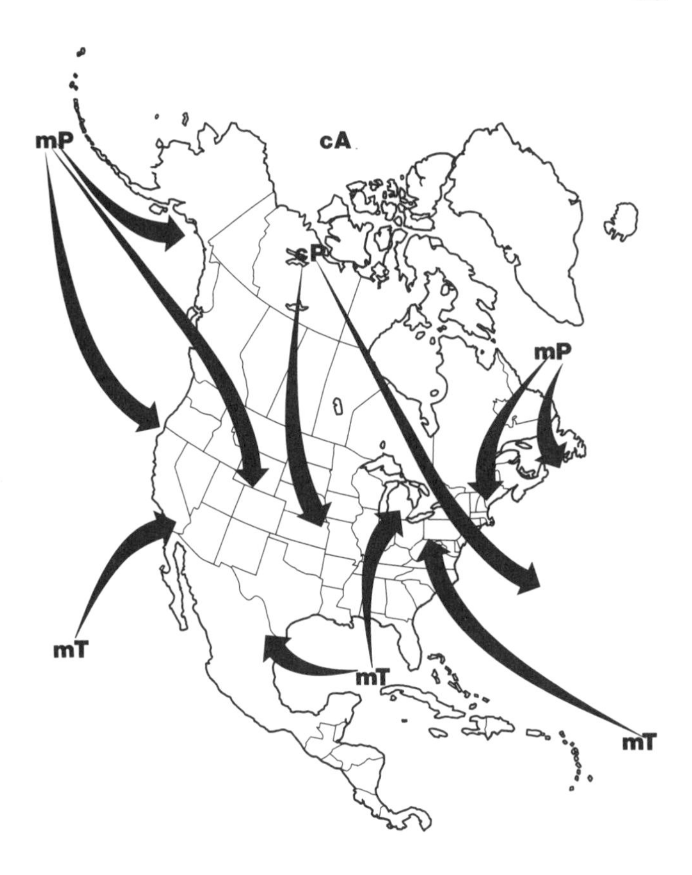

Fig. 8-1 Sources and general paths of air masses in North America (mP—maritime polar, cA—continental arctic; cP—continental polar, mT—maritime tropical).

CONTINENTAL POLAR

Continental polar (cP) air masses have their origin over the ice- and snow-covered interior regions of the continents. Due to the long winter nights, the earth in these regions loses more heat than is replaced through radiation by the sun. The lower atmosphere near the earth's surface becomes colder than the higher atmosphere, creating a condition known as a *temperature inversion.* In the United States, these cold air masses, born as they are in interior regions of the far north, enter somewhere east of the Rocky Mountains. They can dip far south, bringing cold, dry weather to much of the central and eastern part of the country. Unseasonal spring and fall frosts usually are caused by the intrusion of a polar air mass over the region.

When a continental polar air mass moves over a warmer area, such as the Great Lakes, it picks up both heat and humidity from the water, becoming increasingly unstable and finally bringing snowstorms to the eastern and southern lakeshores, where it again reaches the cold land.

In the summer months, when the surface of the polar regions warms, there is no longer an inversion because the atmosphere is warmed from below and, as it passes in a southeasterly direction over land, it brings welcome relief from the hot spells of summer.

MARITIME POLAR

As implied by the name, *maritime polar* (mP) air masses originate over the oceans of the polar regions. This type is both cold and humid. Maritime polar air born in the north Pacific influences the weather along the west coast of the United States. Since the polar oceans are considerably warmer than the polar interior regions, mP air is not nearly as cold as cP air, so the region west of the Rocky Mountains enjoys a milder climate than the rest of the country.

Some cP air masses reach the west coast from Siberia, but since these must travel over the Pacific, they are warmed as they proceed on their southeastward journey. Along the way, they also absorb moisture from the ocean and change into mP masses, which frequently bring clouds and precipitation to the first land they reach.

Since the general circulation around the earth is eastward, polar air masses originating in the Atlantic have a greater effect on European weather than on U.S. weather.

CONTINENTAL TROPICAL

In the summer, *continental tropical* (cT) air masses originate over subtropical deserts and steppes. The air is hot and dry, producing searing days and cold nights. The cT air masses that form in North Africa travel northward, picking up moisture along the way and producing showers and thundershowers in southern and eastern Europe. Since the Western Hemisphere has no large desert regions in the tropics, there is no source for that type of air, and, consequently, the United States is not affected by cT air masses.

MARITIME TROPICAL

Hot, humid *maritime tropical* (mT) air is formed over the warm oceans of the world. It is often unstable and serves to transport heat and moisture to colder regions.

In the summer, mT air from the Gulf of Mexico, the Caribbean, and the adjacent Atlantic enters the United States east of the Rocky Mountains, bringing hot, humid conditions to the central and eastern part of the country. Since this air is heavily laden with moisture, it accounts for much of the precipitation over the affected regions. The closer an area is to the source of the mT air, the greater the rainfall that can be expected.

Tropical air from the Pacific affects the west coast of the United States. It brings clouds but does not generate appreciable precipitation until it has to rise over the western mountains, which then causes release of the moisture on the western slopes.

OTHER AIR MASSES

Two other types of air masses form—one over the arctic basin and the ice cap of Greenland called *continental arctic* (cA), and the other called *maritime equatorial* (mE). Even though cA masses originate over arctic waters, those waters are ice-covered throughout the year, making the air mass dry and very cold. The difference between cA and cP is mostly one of degree, with the former being even colder than the latter. The difference between mE and mT is also one of degree. There is no continental equatorial mass, since equatorial regions are ocean-dominated.

9. Fronts and Storms

During the passage of an air mass, which may cover thousands of square miles, weather remains more or less constant. But since the atmosphere is always in motion, eventually two differing air masses will meet. The line dividing them, perpendicular to the earth's surface (where it actually intersects the surface), is called a *front*. It was so named during World War I because this is where atmospheric battles take place. To differentiate, the more or less horizontal separation between the masses is called the *frontal surface* (fig. 9-1).

Warm fronts, being lighter, will ride up over cold fronts. If the earth did not rotate, the surface of separation between them would be horizontal. It would be the same as when oil and water are poured into a glass jar. They separate, with the oil floating on top of the water. But the earth does rotate, causing the frontal surfaces to be inclined to the horizontal. If the masses were moving at the same speed in the same direction, the separation would be maintained. Generally though, they will move in different directions at different speeds, and the faster one will advance into the other. Mixing then occurs and the weather changes. During the approach of a well-developed warm or cold front, the pressure near the center generally drops. As it does, the front's forward speed increases.

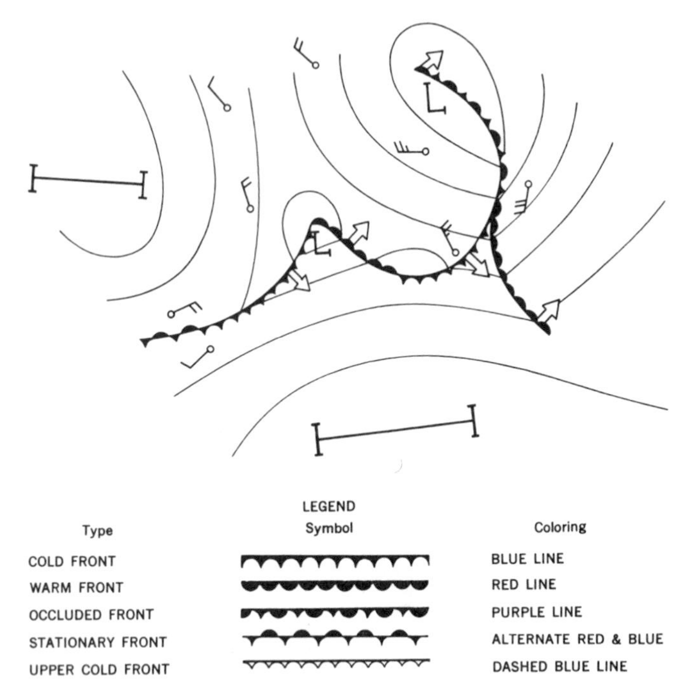

Fig. 9-1 Designations of fronts on weather maps. (Courtesy Bowditch)

WARM FRONTS

When a warm air mass enters an area previously occupied by a cold one, it is called a *warm front.* Friction near the surface is greater than friction aloft, causing the air aloft to move more rapidly than that near the surface and creating a forward slope at higher altitudes. An average slope of a warm front is about 1:200. In other words, 200 miles ahead of the surface location of a warm front, the frontal surface would be 1 mile in altitude. Weather patterns at that height have changed while the surface front is still 200 miles back.

As warm air supplants cold, the pressure falls, indicating the approach of a front. In addition, warm fronts are accompanied by changing cloud formations from cirrus to cirrostratus to altostratus to altocumulus and finally to nimbostratus (see chapter 6). Accompanying precipitation usually starts with drizzle and showers and ends with steady rain when the nimbostratus pass overhead. As the front moves out of the area, the temperature rises, the wind shifts to the right (in the Northern Hemisphere), and the rain stops. Visibility during a front's passage is often limited by fog or mist, which may persist for some time after the front has moved out of the area.

COLD FRONTS

When a cold air mass is the active one, it cuts under the warm air, lifting it to greater heights. A steeper slope is created—somewhere between 1:25 and 1:100—and the slope is in the opposite direction from that of an active warm front. In other words, the warm air is scooped up by the cold air and rises over the top as the colder, denser air moves under it.

Since a *cold front* has a steeper slope and moves with greater speed than a warm front, weather created by its passage tends to be more violent. The degree depends on the temperature and density differences between the masses.

Cold fronts generally are preceded by rain and a cloud sequence of altocumulus or altostratus, followed by nimbostratus, then cumulonimbus. The huge thunderheads that dominate during the front's passage bring heavy rain or hail. As the front passes, the wind shifts abruptly clockwise and can become very squally along this wind-shift line. The pressure rises and the temperature drops (fig. 9-2).

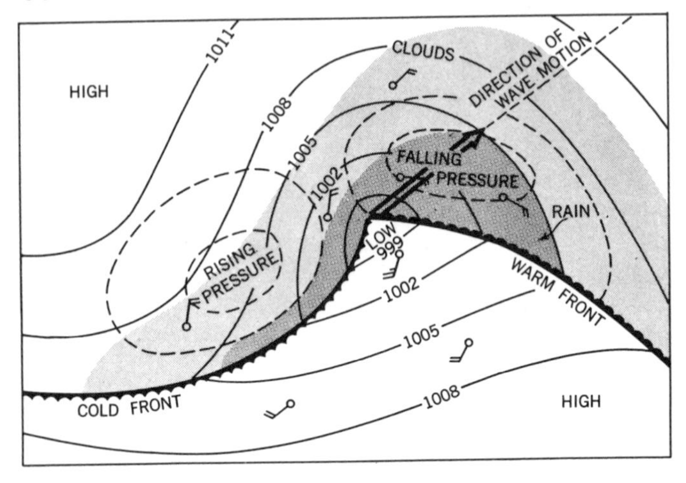

Fig. 9-2 A fully developed frontal wave. (Courtesy Bowditch)

STATIONARY FRONTS

When the surface position of a front does not change, it is called a *stationary front*. Weather conditions in this type generally are akin to those in a warm front, except that they are milder. Sometimes such fronts move back and forth for a short distance, making their future course difficult to predict. They must be watched carefully, as they can develop into strong fronts without much warning.

OCCLUDED FRONTS

An *occluded front* is formed when advancing cold air forces warm air upward and a new front develops between the cold air and the air over which the warm front passes (fig. 9-3).

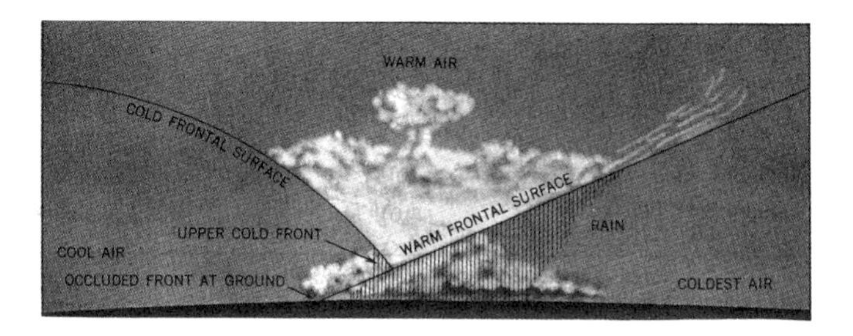

Fig. 9-3 An occluded front. (Courtesy Bowditch)

CYCLONIC DEVELOPMENT

As noted above, the passage of a front causes a drop in atmospheric pressure. This happens particularly when the motion of the two air masses is opposed and the friction between the two causes the frontal line to bend into the shape of a **V.** This bend, or **V**, moves, as do all weather systems, in a generally easterly direction. Warm air rises into the crest, forming a low at the crest or just ahead of it. The colder, denser air circles the crest in a counterclockwise direction, pushing the cold front around until it chases the warm front, overtakes it, and meets the cooler air ahead of the warm front. Together they force the warm air upward, intensifying the low. This process produces an occluded front, with the warm air mass moving aloft. As lifting continues, pressure gradients and winds intensify until all the warm air has been displaced, the front no longer exists, and the pressure gradient again weakens and dissipates the cyclone.

As mentioned in the previous chapter, high winds can occur when the airflow is anticyclonic and the skies are clear, but most severe weather takes place during a period of cyclonic flow.

HURRICANES

All hurricanes are cyclones, but not all cyclones are hurricanes. In fact, very few are. Cyclones bring clouds with or without precipitation and are sometimes, but not always, accompanied by strong winds. Frontal lows can also bring squalls with strong winds of short duration.

Hurricanes are a particular kind of cyclone with whirling winds that are small in circumference (normally between 50 and 125 miles in diameter) and have very steep pressure gradients. A storm is considered hurricane strength when the wind speed reaches 64 knots—force 12 on the Beaufort scale (see chapter 10). Winds of well over 100 knots with higher gusts, frequently develop into mature tropical cyclones with gale-force winds frequently extending as far as 600 miles from the center of the storm.

Hurricanes form in the warm oceans of the world, between latitudes 5° and 20° on both sides of the equator—between the doldrums and the strongest trade winds—late in the summer season, when the water temperature has reached at least 80°F. Interestingly enough, hurricanes do not form within 5° of the equator, the area of the doldrums, probably because the Coriolis force there is too weak to create the necessary rotary motion.

The process begins with small tropical storms in regions of low-level convergence and lifting. Anyone who has visited tropical regions during the summer will recall the dramatic cloud formations and spectacular sunsets that precede the many storms that pass through the tropics then. Most do not develop into full hurricanes. In order for them to do so, the divergence aloft must be much greater than the convergence at low levels, thus fueling and strengthening the surface low. As a budding hurricane moves over large expanses of warm water, it continually picks up moisture (see chapter 5). The greater the wind intensity, the higher the rate of evaporation. The heat lost by the water during evaporation is then released by the atmosphere, making the air ever lighter and increasing the divergence aloft. It also makes the air warmer, which again increases the rate of condensation and spurs more release of heat. As the warm, moist air rises, the pressure continues to fall, the winds increase in velocity, and more air rushes in from areas adjacent to the low. As long as conditions remain the same, a hurricane will continue to perpetuate itself and strengthen until it hits land, where it is slowed by friction and deprived of the moisture it needs, or until it takes a turn away from land and travels to ocean areas, where cooler water and air temperatures gradually rob it of its strength.

EXTRATROPICAL CYCLONES

Hurricane-force winds (force 12) occur frequently during the winter in many oceans of the world. They differ from tropical cyclones in three ways:

1. They form around a cold core rather than a warm one;
2. They cover a much larger area; and
3. The strongest winds generally are found in the storm's periphery, not around its center.

Extratropical cyclones frequently form when cold fronts from the north meet warm, moist air over the oceans in higher latitudes. (The warm front is accompanied by the parade of cloud formations mentioned earlier.) As the front passes, winds veer, pressure rises, and the skies clear. But the fast-moving cold front will not be far behind. As it invades the area, tossing the warm air upward, it creates convergence at low levels and divergence at higher levels, indicated by falling atmospheric pressure and increasing wind speeds. Such storms, accompanied by mountainous waves, may travel for a thousand miles or more without abating.

In the Northern Hemisphere, extratropical cyclones usually originate in the Pacific near China and Japan, from where they travel eastward toward the coasts of Alaska and British Columbia. In the winter, they may also reach as far south as Southern California and affect the entire west coast of the United States. In the Atlantic, a similar belt extends from the eastern part of the United States to Iceland and northwestern Europe. In both cases, the cyclones are more severe in winter, when they can occur farther south than is normal in summertime.

TORNADOES AND WATERSPOUTS

Tornadoes—cyclonic whirlpools of air wound tightly around a center—generally develop in connection with a strong cold front that has produced thunderstorms and squally weather. Over land, their winds may exceed 300 miles per hour, but the center, where the pressure is so low that it is almost a vacuum, creates the greatest damage by causing buildings to explode. *Waterspouts* are tornadoes formed over water, where temperature contrasts are slighter and the resulting wind strengths are mitigated.

Because a spout funnel can be seen a long way off at sea, moves very slowly, and has a diameter of only a few feet, it does not present the danger to mariners that its appearance might suggest. Evasive action usually is possible. It should also be noted that waterspouts are composed mainly of spray and mist. Solid water extends no higher than about ten feet from the spout's base.

THUNDERSTORMS

Thunderstorms are short-lived phenomena produced in clouds of great vertical development and intense updrafts—notably in cumulonimbus. Temperature imbalance in the atmosphere, with warm air near the surface and cold air aloft, creates strong updrafts that carry moisture to the cold upper areas of the clouds. This updraft of warm air reaches levels where it becomes saturated and causes visible droplets to form. As the water vapor changes to liquid or solid ice particles, the resultant release of latent heat creates the energy for further development of the cloud. Eventually, the droplets grow large and heavy enough to fall against the updrafts, causing precipitation. The precipitation, in turn, creates a downdraft more powerful than the original updraft, thus triggering a change in the wind flow within the storm.

The updraft/downdraft combination represents a single storm unit. Any single thunderstorm is small—usually no more than 2 miles in diameter—and short-lived—lasting only about half an hour and not necessarily accompanied by precipitation. On the other hand, such storms may occur in large groups covering hundreds of miles, with torrential rain or hail and ferocious, gusty winds. These are the thunderstorms that spawn tornadoes and waterspouts.

Lightning occurs as a result of the development of an electrical field within the cloud. Normally, a positive charge is found in the upper regions of the cloud, whereas a negative charge is found in the lower regions and near the ground. As the thunderstorm passes over the surface of the earth, a positive charge is induced on the ground. The attraction between the positive and negative charges increases as the storm develops. Since air is a poor conductor of electricity, lightning occurs only when the electrical charges become so strong that they are able to overcome the resistance of the air, thus permitting the current to flow between the charges. Lightning can travel from cloud to cloud—which results in the horizontal lightning often seen in the tropics—or it can travel from cloud to ground or even from high structures on the ground to the cloud (fig. 9-4).

In the midlatitudes, thunderstorms usually form when a cold front pushes warm air aloft. In the tropics, such storms are the result of rising air from converging wind systems. The greatest thunderstorm activity is found in the region of the intertropical convergence zone, where the northeast and southeast trades meet and force the warm, moist tropical air upward. The greater the temperature difference between the surface air and that at successive heights, the greater the instability of the atmosphere and the more violent the storm. Updrafts continue until they enter a stable layer in the troposphere, or, in extreme cases, they enter the stable stratosphere. When the rising air encounters such a barrier, it spreads out, forming the characteristic anvil-shaped cloud.

Tropical cyclones, extratropical cyclones, tornadoes, and thunderstorms are the only violent storms experienced on earth. Fortunately for boaters, they are mostly seasonal and can be expected in specific regions at certain times of the year. But high winds and big waves can occur at any time in any place and present problems for small-boat navigation. In Part II, the emphasis will be on how to forecast dangerous conditions and what to do about them.

Fig. 9-4 Forked lightning showing main branch, which strikes the earth first and is the main channel for the electric discharge. (Courtesy NOAA)

Part II
Weather and the Boater

10. Wind and Water

Wind is, or should be, the prime concern of every boater. How hard will it blow? From what direction? Because waves are created by wind, the significance of wind intensity relates directly to the condition of the boating waters.

A wave has four components:

1. A crest—the highest part;
2. A trough—the lowest part;
3. A length—the distance between successive crests; and
4. A period—the speed of movement of one wave.

Waves develop as a result of wind, but they also are affected by other factors:

1. Fetch—the distance the wind has traveled over the surface of the water;
2. Time—how long the wind has been blowing; and
3. Depth—the distance from the surface of the water to the bottom directly below.

THE BEAUFORT SCALE

When there is no wind, the water is flat. If the wind picks up, small ripples form and proceed in the same direction as the wind. As the wind velocity increases, the ripples become waves, which grow in height commensurate with the wind strength.

Wind strength frequently is given in terms of what is known as the *Beaufort scale,* a numerical system devised in 1805 by Sir Francis Beaufort, an admiral in the British navy. It was designed as a guide to how much canvas a fully rigged warship of those days could carry. The system uses the appearance of the sea surface to estimate wind strength. Even today, careful study of the Beaufort scale relating wind to water conditions is of major importance in enabling any boater to determine how much is too much for a particular boat (figs. 10-1a, b, c).

WIND ON OCEAN WAVES

Once force 7 has been attained, the wind, acting mostly on the crests, pushes the crests faster to leeward than the rest of the wave can travel, causing the waves to break. As the wind increases, the breaking water increases in both quantity and strength. Oceanographers report that pressures in excess of 2,000 pounds per square foot have been measured in breaking waves, whereas wind pressures rarely exceed twelve pounds per square foot. It becomes obvious, therefore, that wind-generated waves can do more damage to a boat than the wind itself.

Be aware that wave height according to the Beaufort scale is not reliable when the wind first starts to blow, because it takes time for waves to build, even though spray and whitecaps will occur rather quickly.

FETCH

Fetch also plays a part in determining the height of waves. The greater the distance over which the wind has blown, the higher the waves will build. But fetch is finite. The

Beaufort Force	*Wind Speed*	*Description*	*Effects at Sea*
0	Under 1K	Calm	The sea is like glass but an uncomfortable swell may be present. Powerboaters love it. Sailors become frustrated and wonder if the wind will ever blow again.
1	1–3K	Light Air	Scalelike ripples appear on the water. Sailors should carry large, light-air sails to enable them to make some headway.
2	4–6K	Light Breeze	Small wavelets form on the sea surface with crests that have a glassy appearance but do not break. Conditions are pleasant.
3	7–10K	Gentle Breeze	Wavelets become larger—up to 2 feet in height. Some crests begin to break. When the wind approaches 10K, whitecaps appear. Sailboats begin to pick up speed.
4	11–16K	Moderate Breeze	Things are beginning to pick up. The wavelets have turned into higher, longer waves, up to about 4 feet. Whitecaps are more abundant. Sailors will change to working sails. Some small boats may begin to experience rough going.

Fig. 10-1a The Beaufort scale (continued on following pages)

Beaufort Force	*Wind Speed*	*Description*	*Effects at Sea*
5	17–21K	Fresh Breeze	Waves, up to about 6 feet in height, take on a more pronounced long form. Whitecaps are everywhere and there is a chance of some spray. Stronger gusts can be expected. Planing powerboats should slow to avoid pounding.
6	22–27K	Strong Breeze	Large waves begin to form. White foam crests are extensive and there is apt to be more spray. Good helmsmanship becomes vital and one must be prepared for stronger gusts and freak waves.
7	28–33K	Moderate Gale	The sea heaps up and white foam from breaking waves begins to blow in streaks along the direction of the wind. Wave height increases to about 14 feet. Planing powerboats should seek shelter quickly.
8	34–40K	Fresh Gale	Wave height increases to 18 feet and the wave length increases as well. The edges of the crests begin to break into spindrift and foam is blown in well-marked streaks. Sailboats reef.

Fig. 10-1b The Beaufort scale

Beaufort Force	Wind Speed	Description	Effects at Sea
9	41–47K	Strong Gale	Waves increase to as much as 23 feet. Dense streaks of foam are present along the direction of the wind. Crests begin to topple, tumble, and roll over. Spray may affect visibility. Tricky steering.
10	48–55K	Whole Gale	Waves are very high, with overhanging crests. Foam is blown in great patches of dense white streaks along the direction of the wind. The tumbling of the sea becomes heavy and shocklike.
11	56–64K	Storm	Waves become so high, to about 40 feet, that smaller ships may at times be hidden from view. The sea is completely covered with long, white patches of foam and the edges of wave crests are blown into froth. Visibility is severely restricted. The wind is howling.
12	Above 64K	Hurricane	The wind whips up the water until the air is completely filled with foam and white driving spray. It is impossible to hear what someone next to you is saying. The voice is blown out to sea before it can be heard. Spray blows across the boat. Everything is wet. Waves are huge.

Fig. 10-1c The Beaufort scale

length of time during which waves continue to grow as a result of fetch depends on the force of the wind (fig. 10-2).

After the wind has died, the wave train continues to move, even though the waves have turned into swells. These can create high surf once they encounter land. This is a good example of how wind in one part of the world can affect sea conditions sometimes as much as a thousand miles away.

Assume now that a large swell is running from the south as a result of a prior wind from that direction. A new wind starts from the southwest, creating a wave train from that direction that is now superimposed on the southerly swells. This sets up an uncomfortable motion in the water, resulting in feelings of *mal de mer* for those aboard the boat. Or, if a strong wind that has been blowing in an area for some time suddenly switches around and picks up from another direction, the two wave trains will be superimposed. When the crests of the two meet, the resulting wave will be much greater in height than either one would be. When a trough from one coincides with the crest from the other, the waters will flatten briefly. The result is a confused sea with (if the winds have been strong enough) some mountainous waves (fig. 10-3).

WAVES NEAR SHORE

On the open ocean, where water depths are great, waves generally are far apart. There is plenty of space for a boat to slide down the crest of one wave and mount the next,

---

Fig. 10-2 (Opposite) The relationship of wind strength (Beaufort number), time (in hours), wave height (in feet), period (in seconds), and fetch (in nautical miles). Note that fetch is finite. (Courtesy Bowditch)

Fetch	BEAUFORT NUMBER																										
	3			4			5			6			7			8			9			10			11		
	T	H	P	T	H	P	T	H	P	T	H	P	T	H	P	T	H	P	T	H	P	T	H	P	T	H	P
10	4.4	1.8	2.1	3.7	2.6	2.4	3.2	3.5	2.8	2.7	5.0	3.1	2.5	6.0	3.4	2.3	7.3	3.9	2.0	8.0	4.1	1.9	10.0	4.2	1.8	10.0	5.0
20	7.1	2.0	2.5	6.2	3.2	2.9	5.4	4.9	3.3	4.7	7.0	3.8	4.2	8.6	4.3	3.9	10.0	4.4	3.5	12.0	5.0	3.2	14.0	5.2	3.0	16.0	5.9
30	9.8	2.0	2.8	8.3	3.8	3.3	7.2	5.8	3.7	6.2	8.0	4.2	5.8	10.0	4.6	5.2	12.1	5.0	4.7	15.8	5.5	4.4	18.0	6.0	4.1	19.8	6.3
40	12.0	2.0	3.0	10.3	3.9	3.6	8.9	6.2	4.1	7.8	9.0	4.6	7.1	11.2	4.9	6.5	14.0	5.4	5.8	17.7	5.9	5.4	21.0	6.3	5.1	22.5	6.7
50	14.0	2.0	3.2	12.4	4.0	3.8	11.0	6.5	4.4	9.1	9.8	4.8	8.4	12.2	5.2	7.7	15.7	5.6	6.9	19.8	6.3	6.4	23.0	6.7	6.1	25.0	7.1
60	16.0	2.0	3.5	14.0	4.0	4.0	12.0	6.8	4.6	10.2	10.3	5.1	9.6	13.2	5.5	8.7	17.0	6.0	8.0	21.0	6.5	7.4	25.0	7.0	7.0	27.5	7.5
70	18.0	2.0	3.7	15.8	4.0	4.1	13.5	7.0	4.8	11.9	10.8	5.4	10.5	13.9	5.7	9.9	18.0	6.4	9.0	22.5	6.8	8.3	26.5	7.3	7.8	29.5	7.7
80	20.0	2.0	3.8	17.0	4.0	4.2	15.0	7.2	4.9	13.0	11.0	5.6	12.0	14.5	6.0	11.0	18.9	6.6	10.0	24.0	7.1	9.3	28.0	7.7	8.6	31.5	7.9
90	23.6	2.0	3.9	18.8	4.0	4.3	16.5	7.3	5.1	14.1	11.2	5.8	13.0	15.0	6.3	12.0	20.0	6.7	11.0	25.0	7.2	10.2	30.0	7.9	9.5	34.0	8.2
100	27.1	2.0	4.0	20.0	4.0	4.4	17.5	7.3	5.3	15.1	11.4	6.0	14.0	15.5	6.5	12.8	20.5	6.9	11.9	26.5	7.6	11.0	32.0	8.1	10.3	35.0	8.5
120	31.1	2.0	4.2	22.4	4.1	4.7	20.0	7.8	5.4	17.0	11.7	6.2	15.9	16.0	6.7	14.5	21.5	7.3	13.1	27.5	7.9	12.3	33.5	8.4	11.5	37.5	8.8
140	36.6	2.0	4.5	25.8	4.2	4.9	22.5	7.9	5.8	19.1	11.9	6.4	17.6	16.2	7.0	16.0	22.0	7.6	14.8	29.0	8.3	13.9	35.5	8.8	13.0	40.0	9.2
160	43.2	2.0	4.9	28.4	4.2	5.2	24.3	7.9	6.0	21.1	12.0	6.6	19.5	16.5	7.3	18.0	23.0	8.0	16.4	30.5	8.7	15.1	37.0	9.1	14.5	42.5	9.6
180	50.0	2.0	4.9	30.9	4.3	5.4	27.0	8.0	6.2	23.1	12.1	6.8	21.3	17.0	7.5	19.9	23.5	8.3	18.0	31.5	9.0	16.5	38.5	9.5	16.0	44.5	10.0
200				33.5	4.3	5.6	29.0	8.0	6.4	25.4	12.2	7.1	23.1	17.5	7.7	21.5	23.5	8.5	19.3	32.5	9.2	18.1	40.0	9.8	17.1	46.0	10.3
220				36.5	4.4	5.8	31.1	8.0	6.6	27.2	12.3	7.2	25.0	17.9	8.0	22.9	24.0	8.8	20.9	34.0	9.6	19.1	41.5	10.1	18.2	47.5	10.6
240				39.2	4.4	5.9	33.1	8.0	6.8	29.0	12.4	7.3	26.8	17.9	8.2	24.4	24.5	9.0	22.0	34.5	9.8	20.5	43.0	10.3	19.5	49.0	10.8
260				41.9	4.4	6.0	34.9	8.0	6.9	30.5	12.6	7.5	28.0	18.0	8.4	26.0	25.0	9.2	23.5	34.5	10.0	21.8	44.0	10.6	20.9	50.5	11.1
280				44.5	4.4	6.2	36.8	8.0	7.0	32.4	12.9	7.8	29.5	18.0	8.5	27.7	25.0	9.4	25.0	35.0	10.2	23.0	45.0	10.9	22.0	51.5	11.3
300				47.0	4.4	6.3	38.5	8.0	7.1	34.1	13.1	8.0	31.5	18.0	8.7	29.0	25.0	9.5	26.3	35.0	10.4	24.3	45.0	11.1	23.2	53.0	11.6
320							40.5	8.0	7.2	36.0	13.3	8.2	33.0	18.0	8.9	30.2	25.0	9.6	27.6	35.5	10.6	25.5	45.5	11.2	24.5	54.0	11.8
340							42.4	8.0	7.3	37.6	13.4	8.3	34.2	18.0	9.0	31.6	25.0	9.8	29.0	36.0	10.8	26.7	46.0	11.4	25.5	55.0	12.0
360							44.2	8.0	7.4	38.8	13.4	8.4	35.7	18.1	9.1	33.0	25.0	9.9	30.0	36.5	10.9	27.7	46.5	11.6	26.6	55.0	12.2
380							46.1	8.0	7.5	40.2	13.5	8.5	37.1	18.2	9.3	34.2	25.5	10.0	31.3	37.0	11.1	29.1	47.0	11.8	27.7	55.5	12.4
400							48.0	8.0	7.7	42.2	13.5	8.6	38.8	18.4	9.5	35.6	26.0	10.2	32.5	37.0	11.2	30.2	47.5	12.0	28.9	56.0	12.6
420							50.0	8.0	7.8	43.5	13.6	8.7	40.0	18.7	9.6	36.9	26.5	10.3	33.7	37.5	11.4	31.5	47.5	12.2	29.6	56.5	12.7
440							52.0	8.0	7.9	44.7	13.7	8.8	41.3	18.8	9.7	38.1	27.0	10.4	34.8	37.5	11.5	32.5	48.0	12.3	30.9	57.0	12.9
460							54.0	8.0	8.0	46.2	13.7	8.9	42.8	19.0	9.8	39.5	27.5	10.6	36.0	37.5	11.7	33.5	48.5	12.5	31.8	57.5	13.1
480							56.0	8.0	8.1	47.8	13.7	9.0	44.0	19.0	9.9	41.0	27.5	10.8	37.0	37.5	11.8	34.5	49.0	12.6	32.7	57.5	13.2
500							58.0	8.0	8.2	49.2	13.8	9.1	45.5	19.1	10.1	42.1	27.5	10.9	38.3	38.0	11.9	35.5	49.0	12.7	33.9	58.0	13.4
550										53.0	13.8	9.3	48.5	19.5	10.3	44.9	27.5	11.1	41.0	38.5	12.2	38.2	50.0	13.0	36.5	59.0	13.7
600										56.3	13.8	9.5	51.8	19.7	10.5	47.7	27.5	11.3	43.6	39.0	12.5	40.3	50.0	13.3	38.7	60.0	14.0
650													55.0	19.8	10.7	50.3	27.5	11.6	46.4	39.5	12.8	43.0	50.0	13.7	41.0	60.0	14.2
700													58.5	19.8	11.0	53.2	27.5	11.8	49.0	40.0	13.1	45.4	50.5	14.0	43.5	60.5	14.5
750																56.2	27.5	12.1	51.0	40.0	13.3	48.0	51.0	14.2	45.8	61.0	14.8
800																59.2	27.5	12.3	53.8	40.0	13.5	50.6	51.5	14.5	47.8	61.5	15.0
850																			56.2	40.0	13.8	52.5	52.0	14.6	50.0	62.0	15.2
900																			58.2	40.0	14.0	54.6	52.0	14.9	52.0	62.5	15.5
950																						57.2	52.0	15.1	54.0	63.0	15.7
1000																						59.3	52.0	15.3	56.3	63.0	16.0

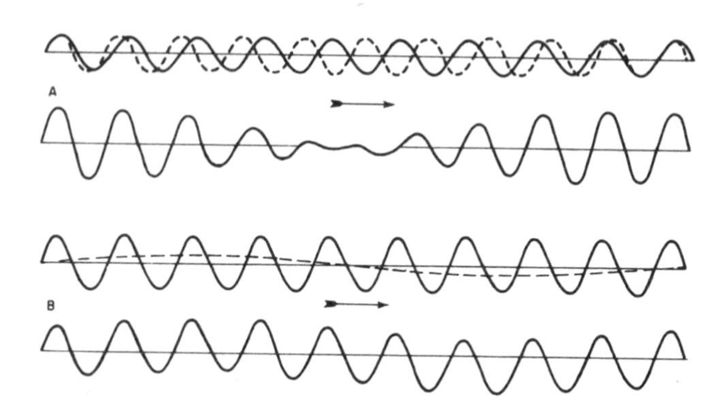

Fig. 10-3 Line 1: Two waves of similar height traveling in the same direction; line 2: The two waves combined; line 3: Wave and swell; line 4: Wave and swell combined. (Courtesy Bowditch)

even in high waves. Unless winds are extreme, there is also enough time between successive crests to navigate the length of each wave. Moreover, winds at sea tend to be fairly steady, blowing from the same direction with little change in intensity for extended periods of time.

Close to a shore, the situation is very different. There the water shallows, slowing the incoming waves by friction with the bottom. As the waves slow, the space between them shortens, as does the period. And as the waves are slowed along the bottom, the crests continue their forward motion and break. But this breaking differs from that of waves at sea, where the pressure is on the tops of the waves. In shallow water, the hindered movement of the lower part of the wave causes the crest to plunge forward with tremendous force. An example of this is waves breaking on a beach, but it can also occur in waters deep enough for boat navigation. The difference is much like that of a man running. If he is pushed on the back of his shoulders, he topples forward, but if his legs are knocked out from under him, he will plunge forward with much greater force (fig. 10-4).

A different situation arises with a coastal configuration of steep cliffs rising vertically from the sea and no shallowing of the waters near the shore. As the incoming waves meet the cliffs, they are turned back, producing a mirror set of waves in the opposite direction. Such seas can produce waves much higher than normal and confused seas, even on a calm day. Add a strong wind and conditions can become downright dangerous. The effects of such seas can be felt more than one-half mile off the coast. This type of sea can also be caused by an exposed breakwater, catching a skipper unaware as he runs for shelter.

Fig. 10-4 Waves breaking on a beach. (Photo by author)

THE 1979 FASTNET RACE

If all this sounds too theoretical, we need only look at the disasters that befell the crews in the Fastnet race of 1979. The Fastnet course starts from Ryde on the Isle of Wight, continues around Fastnet Rock off the southwest coast of Ireland, and ends in Plymouth, England—a distance of about 600 miles. The race takes place in August, when the summer winds usually are light to moderate, but that year a freak storm overtook the fleet—of the 303 boats that started, only eighty-five finished. Twenty-four were abandoned (of which five sank) and fifteen crew members were lost.

On August 11, 1979, at the start of the race, an extratropical cyclone had moved across the United States and was centered over Nova Scotia. It was moving in a generally easterly direction at a speed of 35 to 45 knots. By the morning of August 13, the storm was approaching the Irish Sea. The wind backed to the southwest and the barometer started to drop. The light winds that had been experienced early in the race quickly picked up to force 5 and then force 6. As the low continued to deepen, the fleet soon found itself in force 8 winds from the southwest. Even though the full force of the storm had not yet reached the race area, the wind-lashed waves were making themselves known. Eventually the winds switched to the northwest and stiffened to force 10, setting up a new wave train from that direction. Chaotic seas developed as the two wave trains became superimposed, knocking down several boats.

But the turbulent seas were intensified by the topography of the ocean bottom in that region. The largest waves, moving with the storm at 35 to 45 knots with a period of twelve seconds and a length of about 700 feet, suddenly hit waters shallowed to less than 300 feet over the many banks on the racecourse. As a result, the waves' period and length shortened, then they steepened and broke with tremendous force. Boats swamped and capsized; people drowned. Those in deeper water fared better.

Such storms, though rare, can happen in any waters. The lesson is that there is safety in deep water, away from shoal areas, where there is ample sea room for a boat to select the most comfortable course and ride out the storm in relative safety. Big seas and shallow waters are always a disastrous combination.

11. Coastal Patterns

Most boaters rarely venture very far from land because of time constraints or because their boats are not suited to the open ocean. But, as discussed in chapter 10, waves near shore tend to be steeper, shorter and frequently more confused than those far offshore. And there are other potential dangers: rocks, reefs, and inhospitable lee shores. Most boaters become familiar with conditions in their local waters, learning the weather patterns and danger signs through experience. Others learn the hard way by running into problems.

For the sake of the safety of the boat and all aboard, it is important to understand the effects of winds and currents in relation to the configuration of the land.

EFFECTS OF LAND

Within 50 miles or so from shore, wind and wave patterns are affected by the land, particularly if that land is mountainous. If the wind is from seaward, it can turn around after it has reached the land and come back with increased speed as it funnels through mountain passes. In such a case, the seas still come from seaward, but the strongest wind comes from the shore, creating steep, short seas and wind from unexpected directions.

A similar situation exists when strong winds from the shore funnel through a low spot in the land. The difference here is that the seas are flat within a mile or so of the shore, since they have no fetch. Both of these cases are examples of the *Venturi effect*—named after the physicist Giovanni Venturi, who formulated the principle that when the flow of a fluid or gas is restricted, it increases its speed of flow.

The Venturi effect is also felt in passes between islands, where wind flow over open waters is suddenly restricted. More air pushes through the opening between the land masses, picking up speed. Such a pass frequently is known as a "hurricane gulch."

THE SEA BREEZE

As explained in chapter 2, wind blows from areas of cooler temperatures and higher pressure to those of warmer temperatures and lower pressures. Since land heats and cools more rapidly than large bodies of water (see chapter 3), certain wind patterns are found consistently in coastal areas. After dark, when the land cools, the wind will blow from the land toward the warmer waters. This wind direction remains until about midday, when the sun has raised the temperature over the land. The waters are now cooler and the atmosphere more dense, reversing the wind direction and bringing a cooling sea breeze to the land. On most days, the sea breeze becomes stronger as the daylight hours pass, particularly on clear days when the sun's rays are most intense. In the absence of other factors, probable wind strength can be predicted from the differences in temperature between land and water. Since nighttime temperature differences are less extreme, coastal winds generally are light or nonexistent after dark.

CURRENTS NEAR SHORE

On the open ocean, waters are unaffected by tides and currents are weak—1 to 2 miles per day. Close to a shore, currents intensify, particularly around headlands, creating a set toward the shore. When accompanied by strong winds from seaward, the result is a dangerous lee shore, and it is essential to give points of land a wide berth.

INLAND WATERS

In inland waters, surrounded by land, the wind switches constantly with no warning. No matter from which direction the prevailing wind may come, it is affected by the land as well as by the temperature difference between land and water. In these landlocked areas, waves cannot build to any

Fig. 11-1 Sailboat race on Puget Sound. Note small wave height in choppy waters. (Photo by author)

great height, but the waters can become choppy and confused, making steering difficult, and a boat can be becalmed one moment and buffeted by high winds the next (fig. 11-1).

In large inland bodies of water, such as Washington's Puget Sound, currents intensify particularly in areas where the waters narrow. Tides also become a factor. As the tide is flooding at one end of the sound, it is still ebbing at the other, and vice versa. When the incoming and outgoing tides meet, riptides form. When the winds are strong, a small boat can be caught in a riptide and suffer complete loss of control.

Fortunately, most inland waters offer many sheltered anchorages where one can rest and await more favorable conditions. For anyone cruising or racing in inland waters, tide and current tables are essential, particularly for slower boats that may be unable to make headway against a strong tidal current.

CROSSING THE BAR

Some harbors are located inside the mouth of a river. Rivers carry large amounts of suspended materials into the ocean. The faster the river's flow, the greater the amount of sediment it is able to carry. Where the river nears the ocean, the riverbed usually widens, the flow is slowed, and some of the sediment settles to the bottom. At the river mouth, the current stops and the remaining sediments are released, forming a bar. Over time, such bars have built up to form a sudden shallow area over which vessels must pass to enter or leave the river. Waves and swells break heavily in that area, setting up confused, unpredictable, and powerful seas. Waves will break on a coastal bar even when the waters on both sides of the bar are relatively calm.

When crossing a bar, it is necessary to take into account winds and tides. If the tide is flooding, it counters the oceanward flow of the river, mitigating the effects of the bar. If the tide is ebbing, the river's flow is hastened, resulting in a swift seaward flow that accentuates the rough conditions of the bar. Strong winds from any direction will further increase the danger of the seas across the bar. To minimize the dangers of a bar, it is best to cross at slack tide. As a second choice, cross at flood tide, but *never* attempt to cross when the tide is ebbing and *never* cross when the seas are rough outside the bar.

12. Handling Heavy Weather

Since boats come in a mind-boggling array of lengths, shapes, construction materials, and means of propulsion, what constitutes "heavy weather" is relative to a particular type of boat. It occurs at the point at which the elements seem to take over, tossing the boat about on increasingly angry waters. For anyone caught unprepared, it can be a frightening experience indeed, and a very poor time to begin wondering about the seaworthiness of the boat and how best to handle it (fig. 12-1).

Fig. 12-1 Sailing in heavy weather. (Photo by author)

Every boat is different with respect to its handling characteristics, which depend in large measure on the shape of the underbody, the ratio of underbody to cabin area, design of the superstructure, and weight distribution. The greater the portion below the waterline, the greater the effect of the seas and currents on the boat. Boats with more surface above the water than below will be more affected by wind. In addition, the position and shape of the rudder, and the type of propulsion—single screw or twin screw; left- or right-handed props; inboard, outboard, or stern drive—will dictate how the boat is best kept under control. Becoming familiar with one's boat and carefully assessing its handling qualities is an important part of good seamanship, and what follows here must be interpreted with respect to the type and configuration of one's boat.

Wind and water always exert pressures on a boat. As these elements intensify, those pressures increase dramatically. Generally, boats intended for use in sheltered waters are not built to withstand the abuse they must take in rough seas, where Murphy's Law prevails: Anything that *can* go wrong *will* go wrong.

According to 1987 Coast Guard statistics, 47 percent of all boating accidents and 63 percent of fatalities were weather-related, occurring in waters ranging from choppy to very rough. It seems appropriate, therefore, to consider how and why these accidents happen and how they can be prevented.

People get into trouble on the water when they do not pay attention to the weather forecasts or to the warning signs in the sky and on the water. Some fail to do so because they become engrossed in a related activity—fishing, for instance. When those big ones are biting, who has time to look at the sky? By the time the seas can no longer be ignored, it is too late to return to a safe harbor ahead of the blow.

Sailboat racers are notorious for ignoring the signs of worsening weather. They want to win, and their competitiveness prevents them from reducing sail or abandoning the course. They assume that boat and gear will survive intact. Usually it does, but sometimes it does not. It is unfortunate that races frequently are called off for lack of wind but rarely, if ever, for too much wind.

Then there is the casual boater who thinks of his powerboat as a second car: He turns on the key and goes whenever the desire strikes. But a boat is not a car. There are no service stations on the water, and calling for help is never a good option. The help may arrive too late. Others may be put at risk. Whenever we venture out on the water, we are on our own. It is up to each skipper to do everything possible to ensure the safety of the vessel, which in turn ensures the safety of those aboard.

The most serious weather-related accidents involve sinkings, capsizes, groundings, gear failures that disable the boat, and drownings.

SINKINGS

Take on enough water and any boat will sink. That danger, of course, is greatest for undecked (open) boats, where water from spray and breaking waves can board easily, particularly if the boat is carrying too many people and thus sitting lower in the water than it was designed to do. As water pours into the boat, more weight is added, which lowers the boat even further, reducing its freeboard and making it easier for successive seas to enter. The added weight also makes a boat sluggish, unable to respond in normal fashion and more difficult to control.

Once a boat is in danger of taking on excessive water, the best defensive action is immediately to select an appropriate course to reduce the amount of water entering the boat. Usually that involves *quartering* the waves—heading the boat into the wind and sea at a slight angle to minimize the steepness of the seas.

There are other ways to retard water entry into a small, open boat. An obvious one is to get rid of it faster than it comes in. That means having a large-enough bucket or bailer aboard to do some good. Another method is to stretch a tarp sternward from the bow, as far back as possible, to deflect at least a portion of the water. That requires some forethought to be sure that the tarp and some line have been stowed aboard.

Nor are larger decked vessels immune to the problem. When a boat is tossed about in heavy seas, anything that is not completely secure will loosen. If the loose object happens to be a hose connected to a through-hull fitting below the waterline, the boat will quickly flood, and bilge pumps may not be able to gain on the incoming water. Proper boat maintenance dictates that all through-hulls be inspected regularly and tested to make certain that they can be closed off easily. Every skipper should also be acquainted with the locations of all through-hulls—head, engine intake, stuffing box, and so on—and have wooden plugs available in the proper size for each opening in case the seacock jams or the gate valve breaks off. If the problem is not eliminated immediately, the boat is likely to be lost.

A bilge pump is standard equipment on boats. In many cases, though, there is only an electric submersible aboard. If the flooding water gets to the batteries, the power system will fail, so it is necessary also to have a high-capacity manual bilge pump aboard as a backup.

There have even been cases where a boat was holed by its own equipment. For instance, anchors carried on deck should never have the rode secured to the hull, even though some boats are provided with a ring attached to the hull for securing the bitter end. Should such an anchor break loose and go overboard, the weight of anchor and rode can tear out the portion of the hull to which it is secured. It is far better to lose an anchor than to damage the hull! Another danger, belowdecks, involves heavy objects, such as batteries. If they are not firmly secured, they can break free in rough seas and break through the hull.

If conditions become bad enough, water will enter wherever it can. Are the cockpit drains large enough to disperse quickly a wave that fills the cockpit? Are the ports or windows strong enough to withstand the pressures? Do all ports and hatches close tightly? One very vulnerable area in many sailboats is the companionway hatch, which frequently is narrower on the bottom than on the top. If the boards are not locked firmly in place, a boarding wave can float them out and overboard, after which there will be no way to keep the seas from rushing below once they have entered the cockpit.

CAPSIZES

Another common weather-related accident is capsize. For displacement vessels, the danger of a capsize is remote except in extreme conditions. But that is not true for small, open boats and planing vessels.

Planing powerboats get much of their stability from motion through the water. If, for any reason, they stop moving, they are at the mercy of the seas, which can capsize them. Therefore, engines must always be kept in top operating condition. But, as becomes obvious to anyone monitoring VHF channel 16, one of the main reasons for loss of engine power is running out of fuel. It takes a lot more fuel to power in heavy seas than in flat ones, and it can take much longer to get where you are going. Leaving the dock without an ample fuel supply can spell disaster.

Boats with tall superstructures frequently have too much windage, which makes them vulnerable to a capsize in heavy weather. Such boats must be particularly careful not to be caught beam-to the wind and seas.

Small, light boats generally capsize because weight was not distributed evenly throughout the boat. Perhaps there was too much on one side, creating excessive heel and giving the seas a chance to heel the boat to the point from which it can not recover. This frequently happens when people try to move about the vessel in heavy weather. It may be necessary to just stay put and hang on. On small sailboats, heel must be countered by adding "people weight" on the high side. In any case, it is a good idea for anyone venturing out in a small boat to practice righting the boat and climbing back aboard when in sheltered calm waters to see just how difficult it would be if it had to be done in a blow.

Large sailboats with adequate weight on the keel to counter the pressure of wind on the sails generally will right before they capsize. Once the rail is under, wind spills from the sail, reducing the pressure. However, multihulled boats do not heel as much as monohulls, cannot spill the wind, and therefore are much more prone to a capsize. For this reason, many multihull sailors add flotation to the top of the mast so the boat cannot turn over completely.

GROUNDINGS

Boats run aground when caught too close to a lee shore, rocks, or reefs. Prevention is simple. Stay well offshore. When running for shelter in bad weather, enter the harbor at an angle perpendicular to the entrance rather than trying to hug the shore before turning in. If a boat is close to land and loses power, the seas, wind, and current will push it onto the shore, the rocks, or the reef. Sailboats, too, should stay well out to avoid being caught on a lee shore, unable to sail out of the situation and not adequately equipped to motor to deeper, safer waters. The shallower the waters, the greater the pressure exerted by the seas against the boat and the more power required to move against it. The only chance of saving a boat in such a predicament is to drop an anchor before hitting the rocks. Anchor, rode, and attachments on the boat must, of course, be strong enough to hold. And the system must be aboard, ready to go should the need arise.

Boats also can run aground in anchorages. This can happen at any time, even in calm conditions, if the anchor is too small to hold the boat, not enough rode is payed out, or the holding ground is poor. At other times, a wind shift may expose the anchorage to wind and seas when previously it had been protected by the surrounding land. If the new wind comes from the land, the dangers are not too great because a boat dragging its anchor will drag out to sea. But if the wind shift comes from the water, a boat dragging anchor will end up on the beach or rocks. It is imperative then, to take good bearings on the land—or, if it is too dark, to keep an eye on the depth sounder. If the boat shows any signs of dragging anchor, the safest option—no matter how unpleasant the prospect may be—is to leave and head out to open waters or find another bay more fully protected from the new wind direction.

GEAR FAILURE

The pounding to which a boat is subjected in heavy seas frequently causes a variety of gear failures. For sailboats, gear failure can mean breaking a rudder, a mast, or rigging. It can mean blowing out a sail. A line washing overboard can tangle in a rotating prop. The steering system can break. Any of these and similar problems can disable a boat, leaving it at the mercy of the seas. Powerboaters should slow their boats to minimize pounding, which can loosen gears, cause structural damage, and break equipment. To prevent gear failure, choose a course that allows the boat to ride as easily as possible and at a speed (particularly for powerboats) that minimizes the heavy pounding a boat can encounter in big seas.

Even if gear failure does not disable the boat, it frequently means someone must go out on deck (or aloft) to alleviate the problem, thereby endangering the crew. A case in point might be a hard dinghy that breaks its deck lashings and is threatening to damage the boat. To minimize the possibility of gear failure, periodically inspect the boat and all its working parts to make sure they are strong enough and in good enough condition to withstand heavy weather. In addition, all boats should have plenty of tools and spares aboard for making needed repairs.

DROWNINGS

Most fatalities result when people end up in the water because the boat sank or capsized or because they were washed overboard. On larger boats, everyone should wear a harness and be attached securely to the boat. And on any boat, everyone should wear a life jacket when caught in heavy weather.

It used to be assumed that those who died in the water had drowned. In the cases of those known to be strong swimmers, that was surprising if the accident occurred close to shore. But in recent years, we have learned that many people in the water die of hypothermia. Even in the summer, when air temperatures are warm, water temperatures are much colder and quickly rob the body of the heat it needs to sustain life. When a person swims or struggles, blood flow to the arms and legs increases. The blood, cooled by the water, circulates through the body and cools the vital organs. Life jackets enable the person in the water to remain motionless, thus conserving body heat. Particularly in higher latitudes (those above approximately 40°), water temperatures are 50°F or lower, and, depending on body weight and physical condition, survival time for a swimmer may be as short as fifteen minutes. In warmer water, that time can be considerably longer. In any case, people must be retrieved as quickly as possible. In good weather and calm waters, it is not difficult to spot a person in the water. In a big sea, it may be very difficult. Thus, every precaution must be taken so that no one ends up in the water, and every skipper should give thought to man-overboard procedures and have a plan worked out in case of such an emergency.

Boating accidents are avoidable. They happen to those who go out on the water unprepared and unaware. Paying attention to the weather is the first line of defense against becoming just another statistic.

13. Forecasting Aids

"Watching the weather" means taking advantage of all available weather information in order to anticipate as accurately as possible the type of weather we will find when we go out on the water. Although we are interested chiefly in the conditions in that tiny speck on the globe that encompasses our local waters, we must start with the larger picture of the motion of the atmosphere and then refine it to apply it more specifically to our own area. That involves tracking approaching weather systems, obtaining information on present conditions over the waters, noting changes in barometric pressure, and looking at the sky for the progression of clouds overhead.

Fortunately, this is possible without having to be a meteorologist. After all, there are many professionals who can do most of the work for us. We watch the results on television newscasts, hear the reports on radio, and see weather maps in the newspapers. In addition, detailed marine weather forecasts are broadcast twenty-four hours a day by the National Weather Service.

The National Weather Service (NWS) is a branch of the National Oceanographic and Atmospheric Administration (NOAA), this country's seventh uniformed service. NOAA is active in the fields of oceanography and meteorology, among others, and has done much to improve our understanding of the forces that shape the earth's weather and climate. Information gathered from surface and upper-air observing stations, radar sites, ocean data buoys, environmental satellites, weather-reporting ships, and volunteer observers is incorporated into daily models of the atmosphere from which weather forecasts are compiled (figs. 13-1, 13-2).

Fig. 13-1 A NOAA weather buoy consisting of a tube of sophisticated electronic gear that records wind strength and direction, surface current strength and direction, and air and water temperatures. (Photo by author)

Fig. 13-2 Scientist aboard NOAA ship Discoverer *monitors weather information. (Photo by author)*

WEATHER BROADCASTS

The best overview of weather is available on television newscasts, which take advantage of graphics provided by satellite photographs. When these are put in motion via computers, it is possible to see approaching weather systems as well as the locations of dominant highs and lows. Some television meteorologists are better than others at explaining the systems shown on the screen, so "shop around." Set aside the time to watch the good ones regularly and try to understand what they mean with respect to the various elements discussed in Part I of this book.

Satellite photographs and weather maps provide an overview of the weather for a very large area, so other means must be used to interpret them with respect to a particular cruising area. For instance, a well-defined low may be moving into your general vicinity. It may pass right through your cruising grounds. Or it may just skirt them, perhaps causing high surf conditions. It could also pass well out of the way or disintegrate before it reaches you.

Information specific to local boating areas is provided by the National Weather Service and broadcast round the clock on VHF frequencies. These broadcasts can be picked up on any VHF radio or on the weather channels on multiband receivers (fig. 13-3). Small, inexpensive weather radios are also available for this purpose.

NWS broadcasts are very complete and are updated every few hours. They furnish information on wind strength and direction, wave height, wave period, barometric pressure, and visibility at many local points. They also provide short-term and long-range forecasts.

WEATHER INSTRUMENTS

In addition to weather forecasts prepared by meteorologists, several simple instruments are helpful in providing clues to impending local weather changes.

NOAA RADIO NETWORK
162.550 MHz
162.525 MHz
162.500 MHz
162.475 MHz
162.450 MHz
162.425 MHz
162.400 MHz

Fig. 13-3 NOAA weather-report frequencies.

Barometers. The most important weather instrument for boaters is the barometer, which measures changes in atmospheric pressure. Until almost the seventeenth century, it was thought that air had no weight. But in 1590, Galileo proved that it did, and that changes in that weight bore a relationship to the weather. Some fifty years later, Evangelista Torricelli, one of Galileo's students, invented the first barometer. It consisted of a tube of even diameter, some thirty inches long and closed at one end. The inside was filled with mercury, after which the open end was inserted in a dish of mercury. As the atmospheric pressure increased, it acted on the open dish to force the mercury up in the tube. As the pressure fell, the mercury in the tube also fell.

Today the most common type of barometer is the aneroid. It has a face much like a clock, with markings showing atmospheric pressure—usually in both inches and millibars. It measures pressure by means of the force exerted on a thin metal element connected by means of a combination of levers to a pointer on the dial. The levers magnify the very slight motion of the element so that it can be read precisely. A second pointer can be moved manually over the barometer's pointer so that changes in pressure can be observed easily.

Before mounting the instrument permanently, you will need to adjust it to the proper reading by means of a screw on the back. The current pressure can be obtained from a National Weather Service report. Once most of the existing error has been eliminated, tap the instrument gently to help it adjust itself. Total accuracy is not essential, since any single reading means little. It is the changes that interest us.

Thermometers. Changes in temperature cause changes in atmospheric pressure. Thermometers must be mounted in the open, but shielded from the direct rays of the sun—not very practical on a moving boat. Besides, we can feel changes in temperature and thus do not really need a thermometer to tell us if it is getting warmer or colder.

Anemometers. Once out on the water, it is useful to know both wind speed and direction. The instrument used for speed measurement is an anemometer, which consists of several cups mounted on short vertical arms attached to a shaft that rotates as the wind blows against the cups. On many sailboats, an anemometer is mounted on the masthead and is connected via a cable to a dial in the cockpit, where it can be read. Hand-held anemometers are also available, although they are not as accurate as mast-mounted instruments, since they are held low, where the configuration of the boat affects the wind speed.

In measuring wind speed, it must be remembered that an anemometer measures apparent wind on a moving boat, not actual wind. Assume that a boat is moving at a speed of 10 knots. If there is no wind, the anemometer will indicate a 10-knot wind. If the wind is blowing at 15 knots and the boat is moving directly into it, the reading will be 25 knots. Conversely, if the boat is moving directly *away* from the wind, the reading will be only 5 knots. Actual wind speed can be estimated from the appearance of the water surface by referring to the Beaufort scale (see chapter 10), but an anemometer is helpful in measuring increases or decreases in wind strength that are not observed as easily.

Wind Vanes. The best aid in determining wind direction is a wind vane mounted aloft. It is pivoted on a vertical shaft with more surface on one end than the other so that the pressure of the wind forces the smaller end to point into the wind. Some wind vanes are connected to an indicator to permit continuous readings, but simpler ones merely sit up there, pointing into the wind, and they must be observed directly. Pieces of string attached to the port and starboard shrouds can also be used to do the job. Again, the direction indicated by a wind vane is the apparent direction of the wind, not the actual direction. (Remember that wind direction is always the true direction *from* which the wind is blowing.)

Psychrometers. Another instrument that ought to be in wider use on boats is the psychrometer. It gauges the relative humidity in the air, from which it is possible to ascertain the likelihood of fog. The most common type used on boats is a sling psychrometer (fig. 13-4). This simple, inexpensive instrument consists of two thermometers mounted together, one of which has its bulb covered with muslin. The muslin is moistened thoroughly, after which the instrument is whirled to induce rapid evaporation of the moisture. Then the two thermometers are read. The closer the readings, the greater the possibility of fog.

Weatherfax. In this age of electronics, very sophisticated weather information is available via an instrument known as Weatherfax. It consists of a computer, a single-sideband radio receiver (either separately or in one unit), and a

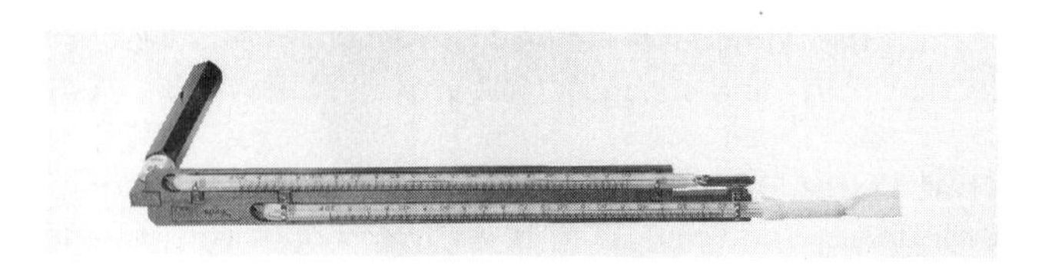

Fig. 13-4 A sling psychrometer. (Courtesy Taylor Instrument Co., Rochester, NY)

printer. By means of this facsimile equipment, it is possible to print out various kinds of daily weather maps aboard the boat. Among the available maps are those showing surface as well as upper-level atmospheric conditions, sea-surface temperatures, and wave heights. Weatherfax equipment is particularly valuable for blue-water voyagers who may not have access to other forecasts.

VISUAL WEATHER WARNINGS

The National Weather Service, when necessary, issues weather warnings for which visual signals are displayed at many marinas, Coast Guard stations, and yacht clubs (fig. 13-5). Some stations display only the day signals, others display both. The NWS considers these signals as supplementary to warnings furnished on their radio broadcasts because they cannot be specific as to time, duration, or direction of a storm.

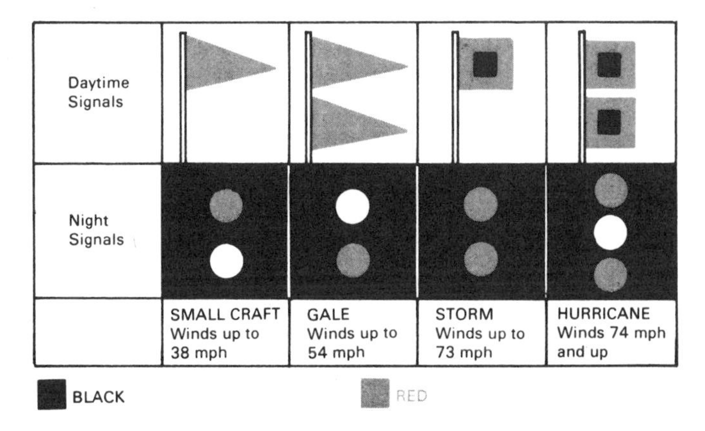

Fig. 13-5 Visual warning signals. (Courtesy U.S. Coast Guard)

We are all familiar with the red pennant that indicates "small craft advisory." According to the Weather Service, the term *small craft* includes small boats, yachts, tugs, barges, boats with little freeboard, or any low-powered craft. This definition covers a diversity of boats and conditions that, though hazardous for some, may be perfectly safe for others. Moreover, such warnings do not differentiate between constant high winds and localized squalls. Sometimes they merely indicate heavy surf conditions. It is up to each boater to decide, on the basis of other information, just what such advisories mean. Nevertheless, a review of the visual signals is appropriate here.

Small Craft Advisory. One red pennant displayed by day, or a red light above a white light by night, indicates winds up to 38 mph and/or dangerous sea conditions.

Gale Warning. Two red pennants displayed by day, or a white light above a red light by night, indicates winds up to 54 mph.

Storm Warning. A single square red flag with black center displayed by day, or two red lights at night, indicates winds above 54 mph.

Hurricane Warning. Two square red flags with black centers displayed by day, or a white light between two red lights at night, indicates that winds over 74 mph are forecast.

Unfortunately, the Weather Service no longer contacts displaying stations when higher winds and/or sea conditions are expected. Most stations now monitor the weather on their own and display the appropriate signals, but this system is not as reliable as the former one.

ENVIRONMENTAL INDICATORS

We have many instruments and extensive information regarding weather at our disposal, but it still is important to heed the more basic signals we can receive by looking skyward and being attuned to the motion in the waters. Weather can change very suddenly, and such changes usually are indicated by the sequence of clouds overhead. We can also see and feel changes on the water. Any blue-water cruiser will attest to the fact that he or she will awaken from sleep when the rhythm of the sea changes—indicating an impending switch in wind direction and/or wind strength. By scanning the waters, it is easy to spot areas of light winds, areas of no wind, and areas of stronger winds. And, as indicated earlier, the waters give a good indication of both wind speed and direction. Taking careful note of the environment at all times is one of the best ways to predict possible changes in weather and sea conditions.

14. Weather Forecasting

The ability to correlate available weather information in order to predict what conditions will be in a particular place at any given time takes some practice and effort, but the more we learn about it, the more interesting and rewarding it becomes. Since it is necessary to track weather over an extended period of time, it is a good idea to begin by taking notes, starting with the large-scale pictures we see on television and working our way down to the local area. Figures 14-1a, b, and c are provided to help make weather watching simpler and more organized.

The two most important aspects to remember are:

1. The atmosphere is in constant motion and travels basically west to east across the country.
2. Changes in temperature and pressure cause the wind to blow. The greater and more rapid the change, the stronger the winds.

LARGE-SCALE WEATHER

The continental United States covers a large area with many types of weather occurring at any given time in different places. The first question to answer is: Where does the weather in my area come from? As was explained in chapter 8, the atmosphere moves in vast homogeneous units known as air masses. The type of conditions they bring depends on where they were formed as well as on

WEATHER LOG 1

SATELLITE PICTURE DATE________________

1. Pressure system over area: High_____ Low____
2. Pressure system approaching: High____ Low____
3. Direction of approach: N____ S____
4. Speed of approach: Fast____ Slow___ Stalled____
5. Clouds: None ____High ____Low ____
6. Precipitation: None____ Light____ Heavy____

PRESENT DAY CONDITIONS:
1. Temperature: High____ Low____
2. Winds: Direction_______ Speed____
3. Clouds: None____ High____ Low____
4. Precipitation: None____ Light____ Heavy____

NEXT DAY FORECAST:
1. Temperature: High____ Low____
2. Winds: Direction_______ Speed____
3. Clouds: None____ High____ Low____
4. Precipitation: None____ Light____ Heavy____

Fig. 14-1a Weather information from television reports.

the route they have taken to reach the local area. If they originated in the polar region, they will bring cold air. If they originated in warm tropical ocean waters, they will bring warm, moist conditions. Whenever two such air masses meet, they form a front, or battle line, which brings unsettled conditions.

Although an air mass moves as a unit, it can intensify or break up, or change its course and speed of travel, depending on the location and intensity of other air masses it meets, or even alter its characteristics due to the ground conditions it encounters (figs. 14-2a, b, c).

WEATHER LOG #2

WEATHER RADIO
DATE____________ TIME____________

PRESENT CONDITIONS:
1. Temperature: ____ Rising____ Falling____
2. Pressure: ____ Rising____ Falling____
3. Relative humidity: ____
4. Waves: Hight____ Length____ Period____
5. Winds: Direction____ Speed____
6. Clouds: None____ High____ Low____
7. Precipitation: None____ Light____ Heavy____
8. Visibility: Good____ Fair____ Poor____

NEXT DAY FORECAST
1. Temperature: Rising____ Falling____
2. Pressure: Rising____ Falling____
3. Wind Direction: Same____ Changing to____
4. Wind Speed: Increasing____ Decreasing____
5. Clouds: Increasing____ Decreasing____
6. Precipitation: Increasing____ Decreasing____
7. Visibility: Increasing____ Decreasing____

Fig. 14-1b Weather information from VHF weather channel.

On the satellite pictures shown on television, cloudy areas are those of low pressure, clear areas indicate highs. When these pictures are put in motion, it is easy to see where the systems have originated and where they are most likely headed. A low entering a particular area can be expected to bring warm air, cloudy skies, and possibly strong winds. A high will bring cooler air and generally clear skies.

Watch the speed at which these systems move into the area. Sometimes they will move rapidly, but other times a

WEATHER LOG #3

PERSONAL OBSERVATIONS
DATE________ TIME________

BEFORE DEPARTURE
1. Pressure: _____ Rising____Falling______
2. Wind Speed: Light____ Moderate____ Strong____
3. Wind Direction: ________
4. Dewpoint Spread: ____
5. Clouds: None_____ High____ Low____
6. Precipitation: None_____ Light____ Heavy____
7. Visibility: Good____ Fair____ Poor____

ON THE WATER
1. Winds: Direction____ Speed____
2. Wave Height: High____ Moderate____ Low____
3. Wave Length: Long____ Moderate____ Short____
4. Wave Direction: Constant____ Confused____
5. Clouds Overhead: None____ High____ Low____
6. Cloud Appearance: Light____ Dark____
7. Clouds Approaching: None____ Light____ Dark______
8. Visibility: Constant____ Clearing____ Worsening____

EFFECTS ON THE BOAT
1. Course steered:_______
2. Engine RPM:________
3. Sail Combination:____________________________
4. Most comfortable course:_______
5. Comments:____________________________________
____
____

Fig. 14-1c Weather information from personal observation.

system may park over an area for several days. Frequently an extensive high will force an incoming low over and around it, sparing the area some inclement weather.

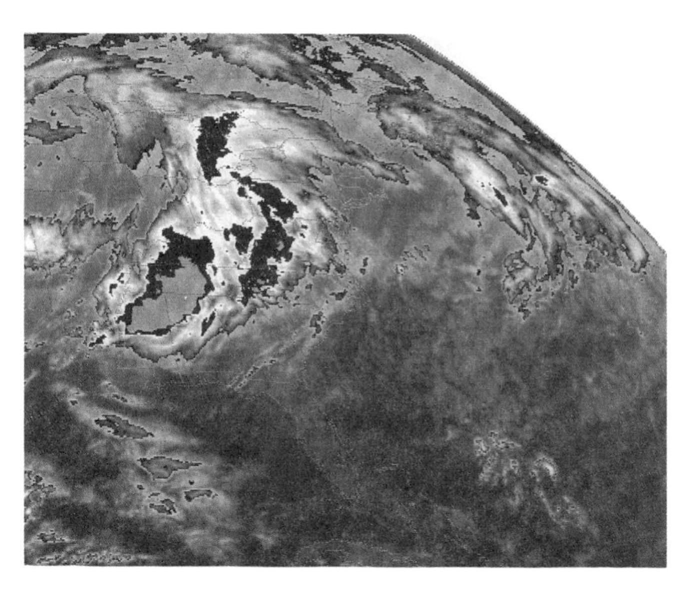

Fig. 14-2a (Opposite, top) Satellite photo of eastern United States, April 2, 1989. A low-pressure system is moving across the Great Lakes area, while the southern part of the country enjoys high pressure and clear skies. (Courtesy NOAA)

Fig. 14-2b (Opposite, bottom) Satellite photo of eastern United States, April 3, 1989. The low has intensified and moved east and south, bringing rain to Tennessee and parts of Kentucky and North Carolina. Another low is behind it, about to pass over Lake Michigan. (Courtesy NOAA)

Fig. 14-2c (Above) Satellite photo of eastern United States, April 4, 1989. The low has moved south and east and now covers much of the region, with many areas of precipitation. (Courtesy NOAA)

Watch also for storms headed in your direction from other parts of the ocean, even if they still are distant. Strong winds as far as a thousand miles away can cause high waves and surf in your local waters.

LOCAL CONDITIONS

After analyzing the present conditions on the satellite pictures, listen to the National Weather Service broadcast. Pay particular attention to wind strength and direction, plus wave height, length, and period. In general, the conditions on the water should relate to what you saw on the satellite photo. However, you must make allowances for the time of day so that you factor in the alternate heating and cooling that normally takes place in coastal areas and gives rise to the afternoon sea breeze.

Whenever you go out on the water, check the report—particularly with respect to wave length and period—in order to get an idea of what constitutes favorable and unfavorable sea conditions for your boat. Short, fast-moving waves may provide a very uncomfortable sea surface, particularly for small boats, even if the waves are not very high.

BAROMETRIC PRESSURE

Many boaters have a barometer aboard, but unless you spend a great deal of time on the boat, it may be preferable to have one at home if you live in the general area of where your boat is located. A barometric reading at any particular time signifies little. What must be determined are the changes, up or down, as well as the rate of change. If the pressure rises or falls rapidly, expect high winds. The more gradual the change, the lighter the winds. In general, a low-pressure system can generate stronger winds than a high.

Changes in pressure are accompanied by changes in temperature. When the air temperature rises, expect the barometer to fall, and vice versa. It is important to remember that here we are talking about air temperature. Air can be quite cool, even though the weather feels warm, particularly on cloudless days when the sun's rays can reach the earth unimpeded.

FOG

Fog is a frightening weather phenomenon that makes the boater feel totally isolated. Frequently it is impossible to see from one end of the boat to the other. Nor is it possible to get one's bearings from the land. Fortunately, it is possible to predict the likelihood of fog by the use of a sling psychrometer (see chapter 13).

Fog is a cloud in contact with the surface. At sea, it frequently occurs when warm, moist air blows over a cooler surface, which cools the air below its dew point (fig. 14-3). As air cools, its relative humidity increases, and eventually it reaches a temperature at which it is saturated and cannot hold more moisture. When this dew point has been reached, dew or fog can form. Therefore, if we know what the dew point is, we can tell whether fog is likely to occur. For instance, after thoroughly wetting the muslin on the wet-bulb thermometer, we twirl the psychrometer until the wet-bulb temperature remains constant, indicating that the moisture will not evaporate further. We can then read both thermometers and, from the difference, enter the table (fig. 14-4) and find the dew point.

Assume that the dry-bulb reading is 76° and the wet-bulb 73°. The difference is 3°. According to the table, the

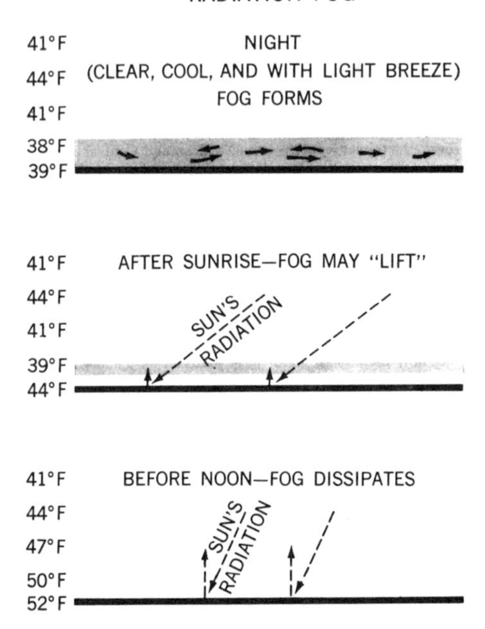

Fig. 14-3 Formation and dissipation of radiation fog. (Courtesy Bowditch)

dew point at that time is 72°. If the air temperature continues to fall, fog is likely to occur. At any time, if the difference between the dry-bulb temperature and the dew point is less than 6° and the air is still cooling, fog can be expected within a few hours.

Boaters in the northern United States are more plagued by fog than those in southern areas. Fog also occurs more frequently along the Pacific coast than the Atlantic coast. Along the south Atlantic coast and in the Gulf of Mexico, fog is virtually nonexistent in the summer and infrequent

Dry Bulb minus Wet Bulb	Dry-Bulb Thermometer Temperature									
	45	50	55	60	65	70	75	80	85	90
1	2	2	2	2	2	1	1	1	1	1
2	4	4	4	3	3	3	3	3	3	3
3	7	6	5	5	5	4	4	4	4	4
4	9	8	7	7	6	6	6	6	5	5
5	11	10	10	9	8	8	7	7	7	7

Fig. 14-4 Schedule of air-temperature dew point.

in the winter and early spring. Along the north Atlantic coast, the preponderance of fog occurs in the summer with the rockbound coast of Maine having the highest number of foggy days. Over the Great Lakes, on the other hand, fog is common from early spring to autumn. The foggy season along the north Pacific coast, down to Northern California, occurs in the summer, whereas in Southern California, fog is more prevalent in the winter months.

CLOUDS

As explained in chapter 6, there are only three basic cloud types—cirrus, cumulus, and stratus—with all others being variations of these. But cloud identification can be confusing, because frequently more than one type is present. Fortunately, it is possible to have an idea of what clouds mean simply by their appearance and development. If they are high and thin, or lower, white, and sparse, expect fair weather. If they thicken and lower, a low-pressure system is moving into the area, possibly with precipitation and/or wind (fig. 14-5).

Basic Cloud Types

Type	*Description*	*Weather Change*	*Precipitation*
Cirrus (Ci)	Thin, feathery ice-crystal clouds that often form long streamers with hooks or curls at the ends.	If in bands resembling mares' tails, may signal approaching storm when followed by Cs or As.	None
Cirrostratus (Cs)	Thin, veil-like clouds that cover the sky totally or partially. May form halo around sun or moon.	Foretell storms if between Ci and As.	None
Cirrocumulus (Cc)	Thin, white layers of clouds with regular arrangement like mackerel's scales.	Usually accompany fair weather.	None
Stratus (St)	Low, gray cloud layer that covers much of the sky. Resembles fog but base does not rest on the ground. Outline of sun may be visible.	No indication of weather change.	Light drizzle possible.
Altostratus (As)	Clouds of bluish-gray, veil-like appearance that obscure the sun.	When they thicken and lower, Ns will form with rain within 6 to 12 hours.	Light rain or snow possible.

Type	*Description*	*Weather Change*	*Precipitation*
Nimbostratus (Ns)	Low, dark, uniform clouds with low, ragged clouds below. Sun is blotted out.	Occur in conjunction with weather fronts.	Steady, heavy rain.
Cumulus (Cu)	Billowy, dome-shaped clouds that form in patches and do not cover the whole sky.	If small, they indicate fair weather. If they lower and thicken, Cb may result.	None
Stratocumulus (St)	Soft, roll-like, gray clouds that tend to merge. Generally dissipate after dark.	Indicate clear, cool evening.	Possible light rain.
Altocumulus (Ac)	A layer of ball-like masses, more or less regularly arranged. They often merge. Show distinct shadows.	When they thicken and lower, showery weather may follow.	None
Cumulonimbus (Cb)	Clouds with intense vertical development that rise to great heights. Tops are frequently in the shape of an anvil.	Occur along strong cold fronts frequently with thunder and lightning.	Heavy rain and possible hail.

Fig. 14-5 Using clouds to determine weather changes.

WEATHER MAPS

Offshore voyagers who have Weatherfax equipment aboard can track weather systems carefully. Both the National Weather Service and the National Ocean Service transmit radiofacsimile charts on a regular schedule. In addition to surface weather charts, these include forecast maps, upper-air (500 millibar) maps, and other maps giving wave conditions, sea-surface temperatures, and frontal analysis. In order to make the best use of Weatherfax, and to find the most favorable conditions and avoid the worst, it is important to learn the meanings of the symbols used (fig. 14-6a, b, c; fig. 14-7).

A careful study of these figures will explain the universal symbols used. Watch particularly for the designation of fronts and the wind arrows. Fronts are illustrated by lines punctuated with a series of semicircles or triangles on the side of the line toward which the front is moving. Wind direction and speed are indicated on both upper level and surface maps as half-arrows flying in the direction of the wind. The feathers designate the wind speed. Each feather represents 10 knots and each half-feather, 5 knots. When wind speed reaches 50 knots, that is shown by a solid triangle in place of the feathers.

An analysis of the movement of different types of fronts combined with wind direction and speed in the general area, provides an excellent guide for forecasting weather related conditions.

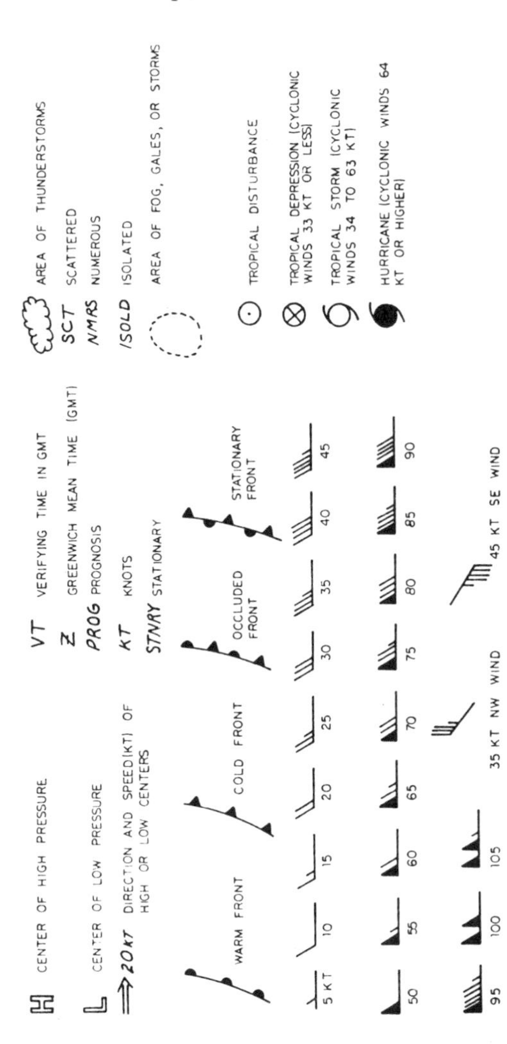

Fig. 14-6a Symbols on weather maps. (Courtesy NOAA)

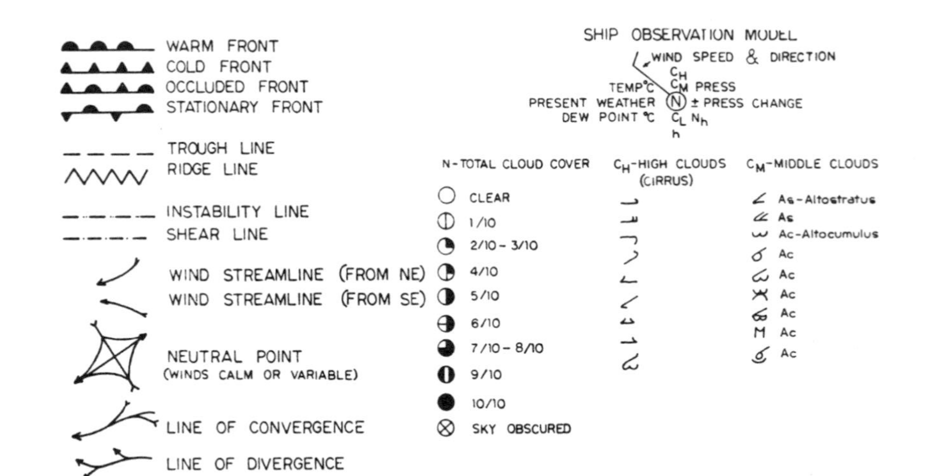

Fig. 14-6b Symbols on weather maps. (Courtesy NOAA)

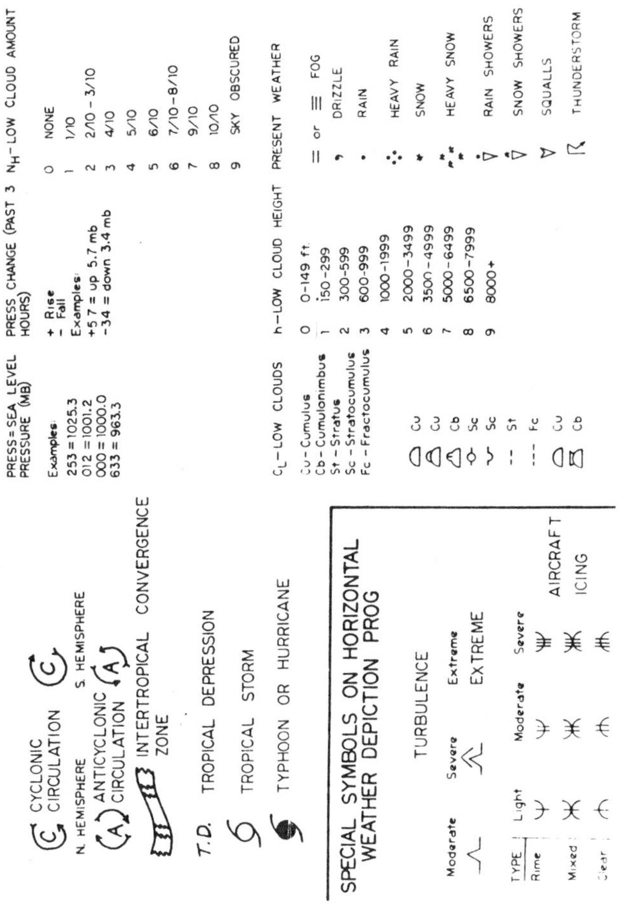

CYCLONIC CIRCULATION

N. HEMISPHERE S. HEMISPHERE

ANTICYCLONIC CIRCULATION

INTERTROPICAL CONVERGENCE ZONE

T.D. TROPICAL DEPRESSION

TROPICAL STORM

TYPHOON OR HURRICANE

PRESS = SEA LEVEL PRESSURE (MB)

Examples:
253 = 1025.3
012 = 1001.2
000 = 1000.0
633 = 963.3

PRESS CHANGE (PAST 3 HOURS)

+ Rise
− Fall

Examples:
+5 7 = up 5.7 mb
−34 = down 3.4 mb

N_H – LOW CLOUD AMOUNT

0	NONE
1	1/10
2	2/10 – 3/10
3	4/10
4	5/10
5	6/10
6	7/10 – 8/10
7	9/10
8	10/10
9	SKY OBSCURED

C_L – LOW CLOUDS

Cu – Cumulus
Cb – Cumulonimbus
St – Stratus
Sc – Stratocumulus
Fc – Fractocumulus

Cu
Cu
Cb
Sc
Sc
St
Fc
Cu
Cb

h – LOW CLOUD HEIGHT

0	0–149 ft.
1	150–299
2	300–599
3	600–999
4	1000–1999
5	2000–3499
6	3500–4999
7	5000–6499
8	6500–7999
9	8000+

PRESENT WEATHER

or FOG
DRIZZLE
RAIN
HEAVY RAIN
SNOW
HEAVY SNOW
RAIN SHOWERS
SNOW SHOWERS
SQUALLS
THUNDERSTORM

SPECIAL SYMBOLS ON HORIZONTAL WEATHER DEPICTION PROG

TURBULENCE

Moderate Severe Extreme

EXTREME

AIRCRAFT ICING

TYPE	Light	Moderate	Severe
Rime			
Mixed			
Clear			

Fig. 14-6c Symbols on weather maps. (Courtesy NOAA)

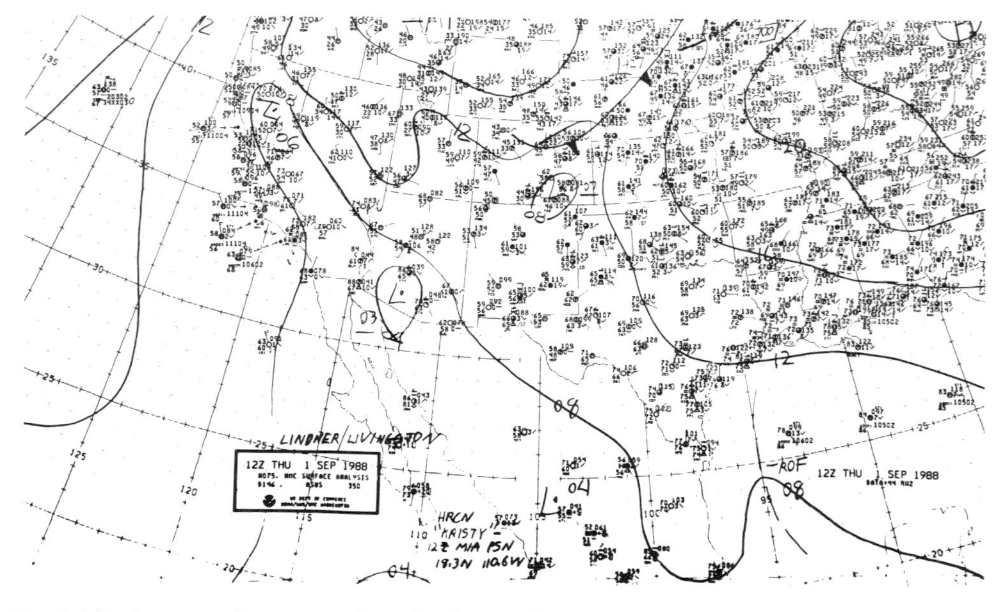

Fig. 14-7 Section of synoptic chart for September 1, 1988. Note hurricane Kristy south of Baja California. (Courtesy NOAA)

15. Hurricanes

There is no weather phenomenon quite so fearsome as a hurricane. These storms can be the most violent, destructive, and awesome meteorological occurrences on the ocean and along coastal areas.

In the United States, the gulf states as well as the southeastern coast sometimes feel the direct fury of a hurricane. How do you safeguard your boat should such a storm threaten? Boaters who weathered hurricane Hugo in 1989 disseminated much information concerning the measures they took and the results of their precautions. There are no easy answers to ensure survival, but general guidelines have emerged from that disaster.

A breakwater cannot stop a forty-foot wall of water when it is lashed ashore by screaming winds (fig. 15-1). Such waves have picked up docks, and the boats tied to them, and deposited the crumpled units in a field far from the ocean. Few docks can withstand the onslaught, and boats remaining in their slips have little chance of survival. Most boats that went through such a storm unscathed sought safety at anchor farther inland. Fortunately, forecasts now give ample warning of an impending hurricane, allowing boaters to find and reach more protected anchorages.

HURRICANE HARBORS

In a hurricane, there is little shelter anywhere from the winds that buffet an entire area. Frequently, however,

Fig. 15-1 A hurricane wave ready to engulf the shore. (Courtesy NOAA)

shelter can be found by sailing up a river or into a lagoon protected by higher ground to seaward in order to diminish the chance of being engulfed by the storm's giant waves (fig. 15-2).

An appropriate shelter during a hurricane—a "hurricane hole"—should have sandy rather than rocky shores. Then, if the boat does run aground, chances are it will not be damaged so badly. The anchorage should also have a sandy bottom that anchors can grab securely. Once there, deploy as many anchors as you have, and, if possible, take a line or two ashore to tie to the base of strong trees. Pay out as much rode as you have available to allow for the expected surge. When the rode consists of all chain, take a bite in the chain and add a long length of nylon line to take up the slack and act as a snubber to ease the shocks. In those cases where line is used for the anchor rode, wrap

Fig. 15-2 A hurricane refuge in a lagoon near Mazatlán, Mexico. (Photo by author)

towels or other material around all parts of the line that might come in contact with the boat to prevent chafe. In addition, it helps to remove everything possible from the topsides—all gear, sails (including the main), dodgers, etc., to reduce windage and prevent damage to the gear.

Some boaters stay aboard, others do not. The decision depends largely on the location of the boat. Personal safety always must be the prime concern. Unfortunately, unattended boats that have not been properly anchored sometimes break loose and create havoc in the fleet—crashing into other boats, causing damage, or even breaking other boats off their anchors. There is no certain way to ensure survival in a hurricane, but anyone living in a region that may at some time be in the path of such a storm would be wise to find out about areas that can provide at least some protection from the wild waters.

HURRICANES AT SEA

What happens to boats caught at sea in a hurricane? One description came from a couple aboard a sixty-five-foot powerboat, a converted aviation rescue vessel, which limped into the anchorage at Mazatlán, Mexico. They had left Guaymas in fair weather without checking any forecast, expecting a few hours' run to Mazatlán. Shortly after their departure, they ran into a hurricane that held them in a fierce grip for two days, during which time they were pushed backward across the water. The house separated from the deck and everything spilled out of the lockers and lay strewn and broken in several inches of water on the cabin sole. They attributed their survival to the fact that, for no particular reason, they had topped their diesel tanks before taking off and were able to run the engines the entire time to hold the boat into wind and seas. Without power, they would have lost stability and been at the mercy of the waves.

Despite all hell breaking loose around them, the couple had not been afraid for very long. The wheelhouse protected them from the weather, no one had to go out in the storm, and they had confidence in their boat. For a time, they said, the waves were so gigantic that they would look up to see fish swimming in the water high above the bridge of the boat. When the wind picked up to over 100 knots, the sea actually flattened: The fierce wind shattered the waves, blowing off the crests. They lost their dinghy off its davits, and a heavy anchor that they thought had been lashed securely on deck tore free and whipped overboard with 300 feet of chain. They were thankful that they had not secured the bitter end of the chain to the ring provided for that purpose in the anchor locker. The force of the ground tackle going overboard surely would have torn out a part of the forward hull. But the couple survived without injury and without damage to the integrity of the vessel.

Because boats are relatively light, they manage to ride the tops of the waves and are not subject to the brunt of the seas. Sometimes they can survive a storm better than large, heavily laden vessels. Nevertheless, it is possible (and better) to heed the warning signs and try to escape the full force of a hurricane.

Tropical cyclones occur during certain seasons. Those originating in the Caribbean have the potential to affect the mainland of the United States. Others in the eastern North Pacific buffet the west coast of Mexico but only indirectly affect the California coast. Occasionally, one will make a direct hit on Hawaii (fig. 15-3). The Caribbean and Pacific hurricanes occur between the months of June and November, generally reaching their peak during Septem-

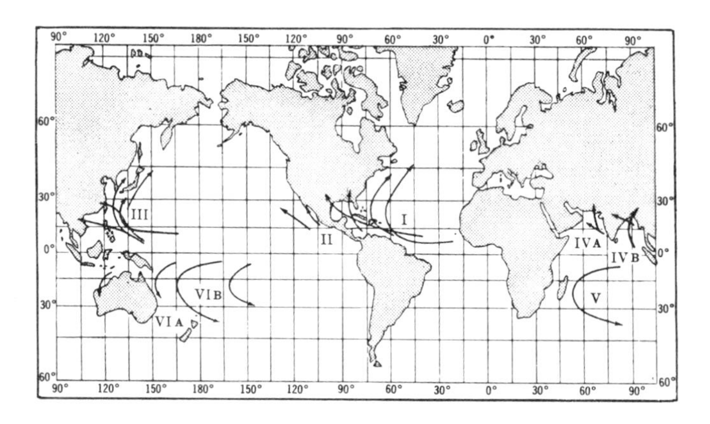

Fig. 15-3 Tropical cyclone areas of the world. (Courtesy Bowditch)

ber. Anyone cruising those areas during the hurricane season is well advised to check the weather forecasts regularly and keep an eye out for warning signs in the sky and on the water.

If a boat is in an area where weather forecasts are not readily available, or are transmitted in a foreign language, it is always possible, with the aid of a short-wave receiver, to pick up the storm warnings that are broadcast every hour on the National Bureau of Standards radio stations WWV and WWVH on 2.5, 5, 10, 15 and 20 MHz. Normally these stations broadcast time ticks continuously for twenty-four hours a day. WWV, broadcasting from Colorado, gives storm warnings at eight, nine, and ten minutes past every hour. WWVH, broadcasting from Hawaii, has the warnings at forty-eight, forty-nine, and fifty minutes past every hour. These storm warnings include general location, latitude and longitude of the storm center, strength and direction of the winds, height of the seas, and the general course and rate of advance of the storm. If a vessel is forewarned in sufficient time, it is possible to track the storm in relation to the boat's position and to follow a course that is likely to lead to safer waters.

There are also warning signs in the sea and the sky that signal the presence of a hurricane in the area. Waves lengthen, but their period shortens. The skies may be clear, but the barometer is restless, vacillating between rising and falling. As the storm approaches, high cirrus clouds in the shape of mares' tails sweep the sky. The barometer begins to fall—gradually at first, then more rapidly. Clouds thicken and lower and rain begins. Great thunderclouds appear on the horizon. The wind intensifies and becomes gusty. The seas build ever higher, torrential rains sweep across the waters, and the wind shrieks. The storm is still several hundred miles away.

As the thunderclouds approach, heavy squalls occur in rapid succession, accompanied by increasingly heavy downpours. Winds intensify. Once the storm is overhead, darkness blankets the area, severely restricting visibility. Squalls become continuous. As the barometer dives, the winds increase in ferocity. Mountainous waves rush in all directions. The sea becomes a frenzied caldron, with biting sheets of spray mingling with the torrential rains and filling the air with water.

If the eye of the hurricane passes overhead, the sky clears for a time and the winds drop to a breeze, but the seas are still hurled in all directions. After the eye has passed, the fury of the wind resumes as suddenly as it had abated. But now it blows from the opposite direction and the cloud sequence is reversed. This rear section of the storm passes more quickly, as it usually is smaller than the front.

Hurricanes are tight, circular cyclones with steep pressure gradients (fig. 15-4). In the Northern Hemisphere, they whirl counterclockwise as they move generally in a northeasterly direction. This means that the right semicircle of the storm moves in the same direction as the

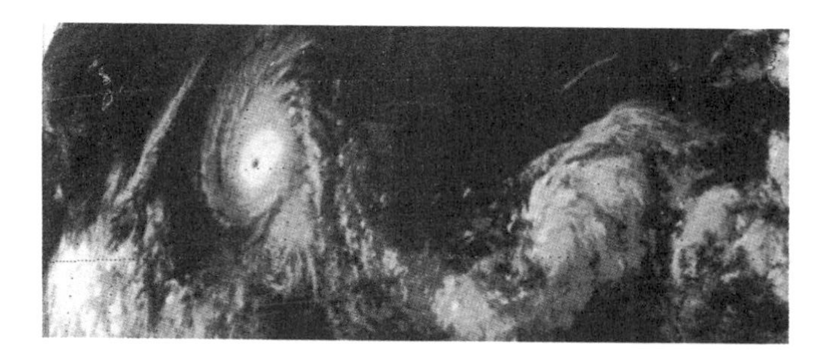

Fig. 15-4 Satellite picture of a hurricane. Note the clearly defined eye. (Courtesy NOAA)

storm itself, and the wind speed is increased by the forward movement of the storm. This section is known as the "dangerous semicircle." In the left side of the storm, the winds blow in a direction opposite to its advance, thus mitigating its intensity somewhat. The left side, therefore, is the "navigable semicircle," and boats should attempt to reach that section of the storm.

In order to find the navigable semicircle, it is necessary to locate the center of the storm. Clues come from the cirrus clouds: The area of convergence of the mares' tails points to the storm's center. The long swells also originate from the center—at least in deep water, where they are unaffected by the configuration of the bottom. In addition, according to Buys Ballot's law (see chapter 7) when an observer stands back to the wind, the low (center of the hurricane) will be to the left and slightly behind. It is also possible to track the center by regular plotting of its position as given in the storm warnings on WWV or WWVH.

With these aids, it usually is possible to determine, with some accuracy, the location of the center of the storm. Full-blown hurricanes are relatively small—generally about 50 to 150 miles in diameter. Even on a slow-moving sailboat, there can be time to take evasive action and avoid the worst of the storm. But the seas will be affected over a great distance, and the going will be rough. It is far better to find a "hurricane hole" in which to anchor the boat and hope for the best.

16. Weather Planning for the Offshore Voyager

Voyaging to distant shores aboard one's own boat is surely the most exciting and rewarding method of travel. So much of the world is accessible by sea that the possibilities are endless. A sturdy sailboat can take you anywhere, whereas a powerboat is limited by its fuel range. Whatever the boat, however, once on the ocean it is at the mercy of the wind and seas. What will be the climate at your destination? What will be the wind strength and direction en route? Which harbors will provide shelter in a blow? And—what is the best course for the return voyage?

Climate and weather are determining factors in the success or failure of an offshore voyage. Good seamanship dictates that they be taken into account when planning such a cruise in order to anticipate the conditions that are likely to be encountered. Too many voyagers run into problems because they did not acquaint themselves with the weather conditions en route. To cite one example: A sailboat left Acapulco, Mexico, bound for Hawaii. The expected length of the voyage was about thirty days and the skipper provisioned accordingly. He set a course that took the boat through the doldrums, where the combination of windless days and squally weather lengthened the voyage to ninety days. Food and water ran out and the crew barely survived the passage. Had he known that he was only a few degrees south of the trade-wind belt, he could have sailed north and found steady winds to carry him to Hawaii in good time.

GENERAL CONDITIONS

A starting point for planning an ocean voyage is to consider the world's wind systems in terms of the desired itinerary (see chapter 4). That will provide a general overview of what weather conditions you might expect in regard to wind strength, direction, constancy, precipitation, visibility, and temperature. In the Northern Hemisphere, the most constant winds are the northeast trades. In the horse latitudes, calms of extended duration can be expected, whereas in the area of the westerlies, changes in wind strength and direction are common. And in the doldrums, calms and squally weather with frequent thundershowers prevail.

From the above, it is possible to develop a picture of what the seas might be like. In the trades, the waves are quite high and long and generally from the direction of the wind. In the region of the westerlies, seas can be confused as a result of the changes in wind direction, and, if the winds are strong, waves can be very high.

To gain more precise knowledge of average conditions in particular areas at different times of the year, information is available from Pilot Charts, Sailing Directions, Coast Pilots, and navigational charts. All of these are published by the U.S. government and are intended for use by ships, but they are equally helpful to boaters.

PILOT CHARTS

Pilot Charts are compilations of average conditions encountered over a period of many years in all oceans of the world. They were the brainchild of Matthew Maury, a navy lieutenant who in 1830 was placed in charge of the Depot of Charts and Instruments, a storehouse of navigational tools for issue to navy ships. Maury had experience at sea, having circumnavigated the globe, and he felt it would be helpful to mariners to have better information about ocean winds, currents, and general weather conditions. He began the ambitious project of compiling this information from old logbooks stored at the depot. He also gave ships' captains specially prepared logbooks in which they could keep track of wind and current data along the ships' routes. Maury received thousands of reports, and these, in conjunction with the logbook information he had assembled, formed the basis of the first *Wind and Current Chart of the North Atlantic*, published in 1847. This enabled ships to shorten the time of their sea voyages by helping them seek out courses that would, in all probability, afford them the most favorable conditions.

Over the years, this information from oceans all over the world has been updated constantly and compiled into the Pilot Charts. Now published by the Defense Mapping Agency Hydrographic Center, Pilot Charts cover the following regions: North Atlantic Ocean, North Pacific Ocean, Northern North Atlantic Ocean, South Atlantic Ocean and Central American Waters, and South Pacific and Indian Oceans.

These charts present data on prevailing winds and currents; percentage of gales, calms, and fog; ice boundaries; and much more. Since these conditions change throughout the year, there is a separate chart for each month, based on an average of conditions reported for that month over many years.

Wind Roses. Of major interest on the Pilot Charts are the wind roses, which are printed in blue and consist of a central circle and eight arrows spaced equally around the circle. The number in the center of the circle indicates the percentage of calms, light airs, and variable winds. The arrows fly with the wind, and their lengths indicate the percentages of total observations in which the wind was from (or nearly from) that direction. To keep the arrows to a manageable length, percentages greater than 20 are indicated by a number on the broken arrow shaft. The number of feathers on the end of each arrow indicates the average force of wind on the Beaufort scale (see chapter 10).

Let's assume we are on a voyage from Los Angeles to the Hawaiian islands. By inspecting the Pilot Chart of the Pacific (fig. 16-1) along that route for the month of July, we see that the predominant wind is force 4 and is northeasterly at first, becoming easterly as the islands are approached. Except in the earliest part of the voyage, no calms should be expected. This makes the east-to-west course a very pleasant passage (fig. 16-2). However, the

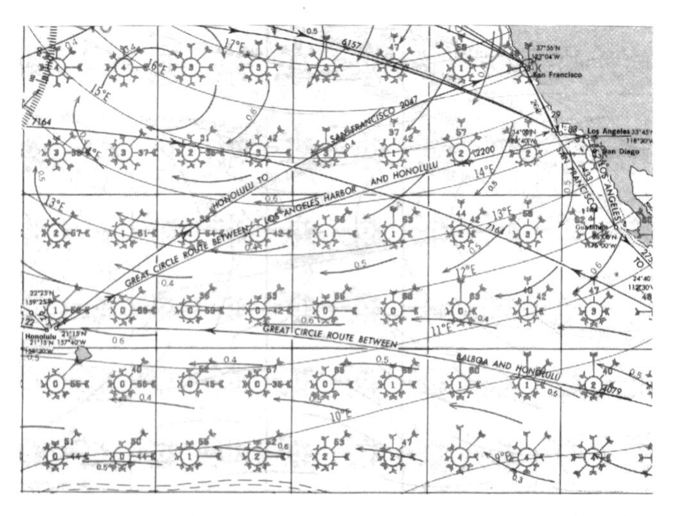

Fig. 16-1 Wind roses and current arrows, from Pilot Chart for July.

return voyage on the same course would be a miserable beat for the entire 2,300 miles.

Surface Currents. The longer arrows, which are printed in green on the chart, show the general surface currents throughout the area. Solid arrows indicate the prevailing direction. Broken arrows indicate probable direction. Wherever currents are significant, a number is given, representing the mean speed per day in nautical miles. For the most part, except in areas of the equatorial counter-current, the current arrows coincide in direction with the predominant winds. If alternate winds were encountered, a change in both direction and velocity of the currents would most likely occur as well.

Fig. 16-2 At the helm in big trade-wind seas. (Photo by author)

On the course of our voyage, the arrows point in the same direction as the wind and show weak currents—only 0.4 to 0.6 knot per day.

Gales and Visibility. To avoid the confusion of too many numbers on the Pilot Chart, a separate small chart insert gives, for every 5-degree square, the percentage of winds in excess of force 8 (over 40 knots) that have been reported (fig. 16-3). In addition, solid blue lines indicate the percentage of observations that have reported visibilities of less than 2 miles.

Temperature. Another small chart insert provides information on average air and sea temperatures throughout the area. Air temperatures are indicated by solid red lines and sea-surface temperatures by solid green lines.

Surface Pressure and Storm Tracks. On this insert, red arrows show the most common tracks of both tropical and extratropical cyclones. Blue lines give the average atmospheric pressure in the region (fig. 16-4).

SAILING DIRECTIONS AND COAST PILOTS

Ever since man began venturing out to sea in boats, there has been a desire to know what lay ahead. What harbors are available? What shelter do they offer? What dangers might be encountered? For centuries, mariners have recorded this data to share with those who were to sail the same course. The earliest of these compilations was the *Periplus Maris Erythraei* ("Circumnavigation of the Erythraean Sea") written between the sixth and fourth centuries B.C. It describes an area of southern Arabia, Ethiopia, and the East African coast and is quite similar to the Sailing Directions we have today.

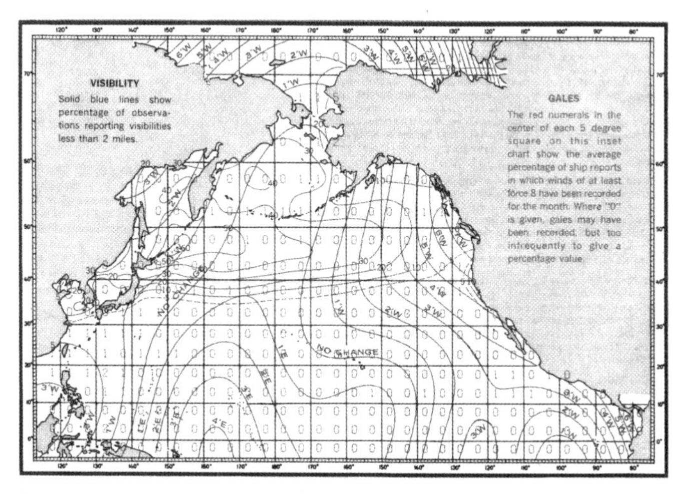

Fig. 16-3 Insert chart of gales and visibility, from Pilot Chart for July.

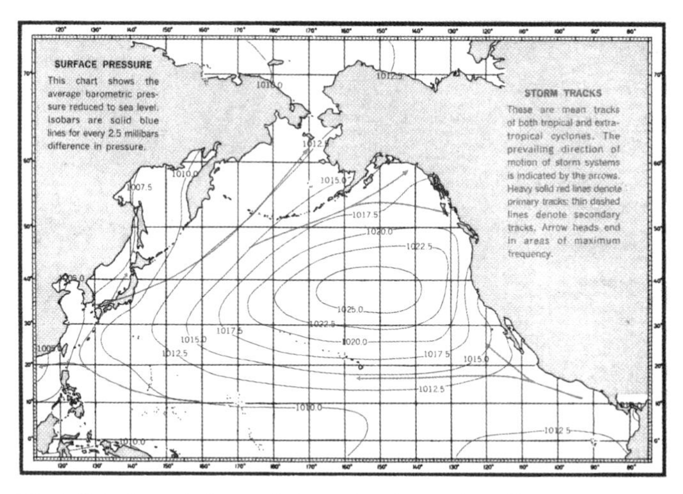

Fig. 16-4 Insert chart of storm tracks and surface pressure, from Pilot Chart for July.

Our modern Sailing Directions are also credited to Lieutenant Matthew Maury. In addition to compiling information with regard to open-ocean conditions for the Pilot Charts, he also was instrumental in gathering data about harbors, shorelines, bays, and so on, which led to the publication of the first volume of Sailing Directions during his service at the Depot of Charts and Instruments.

Sailing Directions are companions to nautical charts. They provide information that cannot be shown on charts, such as detailed descriptions of coastlines, channels, harbors, port facilities, dangers, aids to navigation, currents, tides, and winds. The format now includes eight Planning Guides and thirty-five en route Sailing Directions.

The various volumes of the Coast Pilot cover the coastal areas of the United States, whereas the Sailing Directions cover the coastal regions of other countries. In all other respects, they are the same, describing the entire area in great detail. Of particular interest with respect to weather are the descriptions of every cove and harbor that can provide shelter along the way and suggestions for the best anchorages for specific wind directions. In addition, warnings are included with respect to local and seasonal storms.

SAILING DIRECTIONS PLANNING GUIDES

For the purpose of cruise planning, the Planning Guides are invaluable. The eight volumes cover all of the world's land/sea areas. Each of the guides is divided into five chapters that cover general information about the countries, oceanography and climatology, warning areas, main ocean routes for ships, and the availability of Loran and other electronic navigation reception.

The second chapter of each guide is the most important with respect to weather. For the areas covered, it details information on currents, tides, wave height, climate, and weather during different seasons, as well as special weather phenomena that may occur there. Particularly helpful are the detailed tables that provide information on specific ports for every month with regard to average sea-level pressure, temperature, relative humidity, cloud cover, precipitation, wind speed and direction, days with fog, and days with snow.

There is never any guarantee that particular wind and sea conditions will be found at the time a passage is actually made, but if gales, calms, and unfavorable wind directions are rare in the area, chances are that any bad weather encountered (which undoubtedly will be the case at some time or other) would be of fairly short duration.

On an extended voyage, it sometimes is necessary to cross stormy areas or beat into big seas in order to reach a desired destination. Such courses should not be stumbled upon by accident. They should be set knowingly, with some reflection on the possible consequences.

Index